MARTIN GURDON

First published March 2020

A CIP Catalogue record for this book
is available from the British Library.

ISBN: 978 1 78521 657 2

Library of Congress control no. 2019950543

Published by Haynes Publishing,
Sparkford, Yeovil, Somerset BA22 7JJ
Tel: 01963 440635
Int. tel: +44 1963 440635
Website: www.haynes.com

Printed in Malaysia.

Series Editor: David Allsop.

CONTENTS

Nobody does drama quite like a backyard brood of hens, a bluffer's insight that you can airily share in the certain knowledge that no chicken expert will dare to disagree with you.

SHAKE YOUR TAIL FEATHER

To be a successful chicken-keeping bluffer you need to know who you're bluffing, and why. And to do that you need a comprehensive introduction to the animal, their human keepers, their history and a reason or two why chickens are arguably the most successful avian species of all time. Here's the most compelling: there are an estimated 50 billion chickens in the world (give or take a few billion)* and a growing number of them have embraced domesticity with admirable cunning and determination. Think of the most successful animal species in history (dogs, cats, horses etc) and ask yourself what they have in common. The answer is simple – they have all built meaningful relationships with human beings, and have astutely succeeded in

* According to WorldAtlas.com quoting United Nations 2009 figures.

persuading them that they have an essential role to play in human society. At some point in comparatively recent history, chickens must have noticed this and thought 'Hang on. Why can't we train those gormless bipeds to provide us with food, shelter and warmth as well . . . ?' And as part of the deal, many millions have now done just that; some by producing eggs (known as 'layers'), most by providing meat (the less fortunate, known as 'broilers'); thankfully few to entertain the morons who enjoy watching the abhorrent brand of animal cruelty known as 'cockfighting'; and the lucky remainder for showing off their magnificent plumage (known as 'fancy fowl'). Admittedly, the majority of chickens don't live in the lap of sumptuous back garden luxury, but countless more now enjoy a happier clucking lifestyle than they did a few short decades ago.

Sooner or later, therefore, and especially given the importance of food in our daily lives, the bluffer could be called into action to pretend to be an expert on keeping a personal brood of chickens. The guiding principle of bluffing, and the raison d'être of the *Bluffer's Guides* (now entering their fifth decade of imparting instant wit and wisdom) is that a little knowledge is a desirable thing. Since that is all most of us are ever going to have anyway, we might as well get to know how to spread it thinly but effectively (like Marmite, itself an excellent addition to a chicken salad sandwich).

Careful manipulation of some rudimentary facts will help you bluff your way through with a reasonable degree of nonchalance. This is especially important in these heady days of instant online information. When everyone has Google literally at their fingertips, failure to polish your bluffing techniques will leave you at a social and professional disadvantage.

This book is therefore not a 'how to' guide, but rather a 'how to pretend you know how to' guide. There is a subtle distinction. As the series strapline reads: 'It's not what you know, it's what they think you know.' So come in hard, drop some esoteric facts, make a few opaque allusions, and with any luck your interlocutors will assume that you have the kind of inside knowledge usually confined to the most erudite of experts on all matters chicken-related.

But first it is necessary to try to understand why amateur chicken keeping has suddenly grasped the zeitgeist, and become an essential prerequisite of a successful urban lifestyle. Until a couple of decades ago there were essentially just two types of chicken keeper – poultry farmers and fancy fowl enthusiasts.

The latter came in a variety of shapes and sizes but the standard template tended to involve a degree of portliness, a preference for tweed jackets, a liking for unfashionable hats and a willingness to wear white coats and strut about self-importantly carrying clipboards at chicken beauty contests. These people could tell you what 'candling' was, knew their Barbu d'Uccles from their Plymouth Rocks, were alert to the dangers of vent gleet and able to discuss the minutiae of chicken husbandry in detail. You could see them at shows, peering into cages containing immaculate chickens, speaking to each other in a language of technical terms that was almost Masonic in its impenetrability.

There have always been amateur hen keepers but they were a largely silent minority until, for no obvious reason, the middle classes in the 1990s discovered the joys of owning a few birds, the result of which was an explosion of interest in keeping chickens. Now, according to the British Hen Welfare Trust, hens are the sixth most popular pet in the country.

It would be a lazy generalisation to say that the newcomers were the sort of people who shop at Waitrose, and would prefer to know personally the chickens that produced their organic eggs, but like many lazy generalisations it has a strong ring of truth.

Keeping chickens became fashionable. People who lived in Fulham townhouses and produced designer children called Ludovic and Gethsemane spent up to £5,000 plus on a bespoke henhouse and a trio of shell-shocked ex-battery hens, so that little Ludo and Gethie, when not being tutored to speak Mandarin or play non-contact vegan rugby, could not only learn about life and death, but also the bleak reality of the food chain while looking after their pet chickens. To be honest they probably won't have done too much looking after. Dipping once more into lazy generalisation territory, having a nanny on hand to collect eggs, muck out and take sick chickens to the vet in the family Chelsea tractor was a useful backup.

Actually, ex-battery hens would often be very effective in demonstrating the facts of life and the inevitability of death because they would often expire quite rapidly, although having hopefully seen out their twilight months in chicken comfort, and of course some boldly, and baldly,* went on for years.

Indeed, the rather splendid British Hen Welfare Trust, which as these words are being written had re-homed 737,030 hens (so their numbers will have climbed significantly since, courtesy of the somewhat addictive rolling counter on their website home page), report that the average flock size is five chickens. Not big enough for a smallholding, but large enough for endless drama. And nobody does drama quite like a backyard brood of hens, a bluffer's insight that you can airily share in the certain knowledge that no chicken expert will dare to disagree with you.

And that is where this short but definitive guide comes in. It sets out to conduct you through the main danger zones encountered in discussions about keeping chickens, and to equip you with a vocabulary and evasive technique that will minimise the risk of being rumbled as a bluffer. It will give you a few easy-to-learn hints and methods designed to ensure that you will be accepted as a chicken-keeping aficionado of rare ability and experience. But it will do more. It will provide you with the tools to impress legions of marvelling listeners with your knowledge and insight – without anyone discovering that, before reading it, you didn't know one end of a chicken from the other, or a tail feather from a wheezer.

* One of the unfortunate maladies affecting battery hens is the stress-induced loss of some or all of their feathers.

Plenty of people who were good with wood began selling cute, bijou chicken coops that resembled everything from country cottages to bungalows.

THE CHICKEN-KEEPING REVOLUTION

As the new millennium got under way a proliferation of books on the subject of the bucolic delights of keeping chickens in an urban garden suddenly flooded the publishing world. The author and broadcaster Hugh Fearnley-Whittingstall appeared on TV and interrupted his squirrel stew to show off henhouses that were raised on stilts and accessed by precipitous looking ramps.

This avian loft living wasn't everyone's idea of chicken nirvana, but there were plenty of people who were good with wood who began selling cute, bijou chicken coops that resembled everything from country cottages to bungalows. By this stage lifestyle magazines with glossy paper and glossy contributors were filled with small ads for these birdy mansions, often being sold for mansion-sized prices.

Then there was the 'Omlet Eglu', which was the apex

of middle-class designer chicken houses. Indeed at the time it was unrivalled as the high point in designer hen accommodation. This double-skinned plastic box had a wacky shape, was sold in a range of bright primary colours and had an allegedly fox-proof wire run area.

It first appeared in 2003 and was the brainchild of James Tuthill, William Windham, Johannes Paul and Simon Nicholls, a group of Royal College of Art design students. It was also an early internet sales success, as it was offered on line when a lot of things weren't. In fact it could be bought as a complete starter kit that included bowls, food and even chickens.

Poultry pioneer James Tuthill kept hens at college, built accommodation for them, and when all of them returned home for the holidays, his mother decided that the birds needed more space and better housing. Her son needed an idea for a design project, had decided he wanted to work with his three friends, and having considered and rejected subjects such as domestic lighting, they decided that a modern henhouse was the way to go, despite meeting a fair degree of scepticism.

'You had to pinch yourself sometimes and think: "What am I doing?"' remarked fellow Omleteer William Windham.

Operating out of an Oxfordshire farm complex run by one of the designer's uncles, the henhouse quartet tried out various prototypes on local acquaintances, helped by a small article in the *Banbury Guardian*. Designs included one with a built-in seat to allow owners to sit and look dotingly at their flock.

Nearly two decades on from its launch the Eglu quartet are still together, sell all sorts of animal-related products and no longer do everything themselves. They've flogged henhouses to Hawaiians and Australians and reckon that around 500,000 chickens live in them.

The production version had a folding paddle front door after prototype testing of a guillotine-style door with a small child revealed that the child liked nothing better than to drop the door on to the neck of any chicken that was foolish enough to emerge from the safely of the Eglu.

The partners arranged for the parts to be manufactured by outside suppliers and brought to the farm where they took it in turns to assemble them in an old stable block and thrash up and down the UK's roads in a Mercedes Sprinter van, delivering the finished products.

At the time of writing, nearly two decades on from its launch the Eglu quartet are still together, sell all sorts of animal-related products and no longer do everything themselves. They've flogged henhouses to Hawaiians and Australians and reckon that around 500,000 chickens live in them.

NOT TO BE SNEEZED AT

Since the year 2000 the trend for keeping chickens has ebbed and flowed as hard core hen fans have stuck with their birds, and avian dilettantes having spent vast sums of cash on buying the necessary equipment and the chickens to go with it, found that looking after them actually involved hard work, occasional heartache and got in the way of other aspects of their lives. They often gave up, usually when their birds expired and the novelty wore off.

There was also the threat of bird flu. In 2006 the voguish world of amateur hen keeping took a darker turn when a dead swan found in Scotland was discovered to have had a particularly virulent flu strain called HN51. The fear was that a variation of this could make people ill if it changed and adapted.

Amateur chicken keepers had to keep their hens under cover and find ways to separate them from wild birds as they washed their wellington boots in disinfectant and wondered whether people in chemical suits would be turning up to kill their pets.

The flu virus is widespread in wild birds and every so often crosses into farmed animal populations. From there it can, and in the past has, mutated so that it can infect human beings too.

The Spanish flu pandemic of 1918, which claimed the lives of between 20m and 50m people worldwide, could well have been incubated in birds before it killed people whose immune systems had endured the stress of the First World War.

The infected swan generated plenty of doom-laden newspaper headlines about bird flu ripping through farm birds, including chickens, en route to changing into something that could kill people. The general state of anxiety wasn't helped when chickens at a Norfolk farm tested positive for another strain of bird flu. Other sporadic outbreaks followed, there were mass culls and movement restrictions, amateur chicken keepers had to keep their hens under cover and find ways to separate them from wild birds as they washed their wellington boots in disinfectant and wondered whether people in chemical suits would be turning up to kill their pets.

This put a big dent in the trend of keeping a few birds for fun, although on this occasion an inter-species pandemic did not result. But it took some time for the demand for pet chickens to pick up again, and subsequent bird flu outbreaks have for some people removed the gilt from the chicken-keeping gingerbread on a permanent basis. However, any chicken bluffer can nonetheless reasonably claim that there are a lot more backyard hens two decades into the new millennium than there were at the start of it.

The *Daily Telegraph* has suggested that there are anything up to 750,000 amateur hen keepers in Britain, up from around 50,000 in the late 1990s. How those figures were arrived at isn't clear, as the British Government's Department for Environment, Food and Rural Affairs (DEFRA) doesn't record flocks of less than 40 birds. But they don't sound unreasonable – even if there doesn't seem to be any more recent data. So any self-respecting chicken bluffer can claim that little has changed, since there hasn't been an obvious fall-off of people who give their chickens names and take them to the vet when they're ill.

HOW EASY IS IT TO JOIN IN?

Some of the billions of chickens in the world could be living in your garden (or what's left of it, see below). If not, the idea of them doing so might appeal.

However, the question that needs to be asked is whether the reality lives up to the fluffy-bottomed, egg-rich, contentedly clucking image of a happy hen that's probably drifting around in the aspiring chicken keeper's head? Probably not, and this book will reveal the reasons why the reality can sometimes be somewhat bleak. This doesn't mean, however, that you should necessarily ignore the hen-keeping urge. People get broody too.

Human beings are by nature hunter-gatherers, yet for many of us the thought of active hunting doesn't appeal, whereas gathering eggs is an altogether more attractive and rewarding prospect. What could be simpler? Well, how about not keeping chickens and going to the shops to buy your eggs – purely for simplicity's sake? You might be a seasoned bluffer, capable of dispensing avian *bon mots* with casual assurance, but if you've decided to actually keep chickens, is the person being bluffed really you?

The way to find out is to consider a few hard facts, some of which are more obvious than others. Consider too that these pearls of wisdom can be shared with chicken bluffers and chicken keepers alike.

THE SMALL PRINT

First off, do you live in the right location to have the birds in the first place? It doesn't have to be a rural idyll – there are plenty of urban chicken owners – but if your garden is used by a family of foxes to make a regular trip to tear apart your bin bags and spread their contents all over the street, putting the temptation of some plump and pampered fresh chickens their way might not be a good idea. Like bluffers, foxes are bright, resourceful and like a challenge. If that challenge is proving that a fox-proof chicken run isn't fox-proof at all, awful tragedy can be the inevitable result.

And consider the possibility that you might have a decent-sized garden which is entirely fox free, but discover that you aren't allowed to keep any sort of livestock in it. Any putative chicken bluffer who wants to go native and acquire some birds would do well to check their house deeds to see whether buried in the small print is a clause that forbids such things.

If your garden is used by a family of foxes to make a regular trip to tear apart your bin bags and spread their contents all over the street, putting the temptation of some plump and pampered fresh chickens their way might not be a good idea.

THE BUSINESS END

Consider too what comes out of the back of a chicken. Eggs and compost. The eggs are genuinely welcome, but the other stuff, a combination of urine and faeces, has a more limited appeal. This stuff is pungent, acidic and decidedly unpleasant and chickens produce loads of it. A quartet of birds will evacuate enough guano to murder an average-sized lawn in months, and their habit of scratching and foraging can turn something close to the quality of a bowling green into a wasteland of scrub and dust bowls if left to their own devices.

But as a chicken proto-bluffer you might counter that this stuff is good compost and will do wonders for your flowerbeds and veg patch. This is true but only up to a point. Even a small flock of hens will produce more muck and excrement than a smallholder will ever need and it won't be long before they have bags of the stuff.

Also, neat chicken dung is powerful stuff. Spread

some on a few tomato plants and they'll go potty, but spread too much too often and they'll effectively go 'bang'. Moderation in all things, including distributing chicken filth on your plants, is the watchword here then, and diluting the stuff with other compostable materials is a good way to prevent exploding vegetables. (Please note: your veg won't *actually* explode. This is a mild literary exaggeration for bluffers to deploy if they feel they might be losing their audience.)

ON THE PROWL

Remember too that chickens eat almost anything and like nothing better than probing, pecking, scratching and foraging. They're almost constantly feeding, which means most of the time they're on the go, moving from one edible treat to the next. Bulbs can be wrenched from flower beds, flowers themselves can be trampled and thrust aside, and repeated, surgical strikes will be made on vegetable patches.

Potential chicken keepers should ask themselves whether there really is room for a few birds. Can they make use of a discreet run area, and is the housing bought for them robust and easy to clean? The 'impress-your-friends-with-how-much-you-know-about-henhouse' specifics will be addressed later in this book.

FELLOW TRAVELLERS

Bear in mind that chicken guano is not just a repository for nutrients, it's also a magnet for pathogens, and if birds stay in one place for a long time, the soil beneath their claws will start harbouring this stuff too, resulting in bugs that can make birds ill. So laying down and shovelling up fresh topsoil and materials such as straw should become part of a chicken keeper's routine, as the bluffer can point out. You might also mention wisely that while straw is a fine avian bedding material, hay is not; because it can produce spores that infect a chicken's lungs. Likewise garden centre-style bark chippings, which look pretty to start with, but rapidly rot, releasing all kinds of spores that can entangle around a bird's breathing apparatus, often with fatal consequences.

Bear in mind that chicken guano is not just a repository for nutrients, it's also a magnet for pathogens, and if birds stay in one place for a long time, the soil beneath their claws will start harbouring this stuff too.

For small gardens in particular, some sort of border fencing might well be necessary so that if chickens do enter the garden sanctuary normally reserved for human occupation, they appear as invited guests on their best behaviour rather than marauders hell bent on eating stuff that doesn't belong to them.

TABLE MANNERS

When dispensing chicken-keeping advice the advanced bluffer can tell anyone with a wild bird feeding table that they will need to keep chickens away from it, because it's an absolute grub banquet, and easy pickings for domestic fowl. Wild birds are positively fizzing with diseases, many of which chickens are more than capable of catching. Keeping them away from these areas can be a time-consuming business, and owners will rapidly become accustomed to the creatures' single-minded determination to get what they want.

Some also have a habit of chasing off the feral birds for whom this irresistible smorgasbord was intended; so inter-species neighbouring disputes can be a problem too.

Neighbouring disputes can also apply to chicken keepers. Although they don't want neighbours vetoing their plans to become avian experts, letting them know well in advance about a plan to buy some garden hens (and share some fresh eggs) might avert all sorts of problems. If chicken agnostics feel aggrieved that their neighbours have suddenly filled an adjoining garden with domestic fowl without warning, the more likely the non-believers are to complain about noise, smell and unwanted visitors (if chicken keepers don't get foxes they might well have to contend with rats). Having these intelligent, resourceful, feathery noisily clucking disease vectors as near neighbours is unlikely to appeal greatly to human ones. So bluffers should advise newby chicken keepers to accentuate the positives, and lay the groundwork with neighbours before introducing their new feathery friends to their new home.

FALLING FOWL OF LANDLORDS

If the putative chicken keeper is a tenant and lives in rented accommodation, the bluffer can gravely inform them they should always make sure their landlord permits having birds on the premises. Ignore this and the hen keeper could find the deposits their birds leave on the lawn could cause them to lose their financial equivalents.

MUCKY BUSINESS

Chickens will quickly make an awful mess in their coops, and cleaning these out regularly is a chore that cannot be avoided for various reasons including good avian health, good human health, clean eggs and because the simple reality is that the cleaner the birds are the less they will need expensive medical intervention.

A sick chicken is not 'only a chicken'. For many chicken keepers the birds can arouse sympathy and concern in an owner in much the same way as a cat or dog can. Or, dare say it, a member of the human family.

A good question for a wannabe hen owner is whether or not they are known to the local vet. If not they will be at some point soon if they actually get to the stage of keeping some birds, because inevitably things will go wrong with them – and although chickens blur the line between pets and livestock, owners may find it difficult to ignore problems if one of their flock starts feeling off colour. The wise bluffer will aver that when confronted with the statement 'but it's only a chicken', it is a fact that although chickens rarely bond with people, the reverse is often true. A sick chicken is not 'only a chicken'. For many chicken keepers the birds can arouse sympathy and concern in an owner in much the same way as a cat or dog can. Or, dare say it, a member of the human family.

DEMANDING GUESTS

Chickens should have their water changed regularly and ideally be fed a couple of times a day. This does not always sit comfortably with owners accustomed to going away for the weekend, or taking regular holidays. You wouldn't leave a dog or a cat in the garden for an extended period, you might politely point out, and you can't do the same with a chicken.

Ask aspiring hen owners if they are able to entice or bully a relative, friend, neighbour, or offspring of the above, to look after their birds if they are away. If not you can let them know, with only the smallest hint of gloating that they will never be able to leave their homes again.

The offer of fresh eggs is an excellent incentive to stand-in keepers for stepping into the breach and taking temporary charge, assuming that their birds can be bothered to lay any. However, bluffers should point out that during periods of moulting feathers and growing new ones, broodiness, or if it's cold, or they just don't feel like it, chickens will pack up laying for weeks or months at a time.

KEEPING REGULAR

Then there is the general day in, day out slog of looking after the birds.

A useful way of gauging the level of chicken-keeping commitment is to wait until the weather turns nasty and objects such as dustbins, pensioners and small cars are being blown past a newby hen keeper's window, and they *still* have to dress up like a trawler fisherman and fling themselves into the meteorological melee to get the chickens up, feed and water them. Is this what they actually had in mind?

Come rain, shine, monsoon or the hangover from hell, filling drinkers, shovelling food into bowls and opening henhouses so that their occupants can peer out at their owners in a disbelieving 'do you really expect me to go out in *that*?' sort of way will be a chicken keeper's lot. So will regularly cleaning out the henhouse from which the birds are refusing to budge. In the summer this can be a malodorous, itchy and sweaty process. On an Arctic afternoon in November it can have a Gulag work party quality. Pointing this out can be highly entertaining for a chicken bluffer.

Perhaps you might now be thinking, 'I've bought this book to find out about how to pretend to be an expert on the joys of keeping chickens and it's full of warnings designed to put people off them.'

To a certain extent this is true, but these admonitions are also part of the reality of keeping hens, and, as you can point out, it's far better to quit while you're ahead and henless than finding yourself tending a flock and wishing you weren't.

So far so miserable, but consider: this is just part of the picture, but it isn't the whole picture, and if people can live with the downsides of chicken keeping, then there are plenty of ups to compensate for them.

A HISTORY OF HENS

As a chicken bluffer you will be on safe ground disseminating profundities on the history of chickens, because nobody or nothing is left alive who remembers when they first appeared.

As with all birds in general, chickens are descended from dinosaurs. Consider the similarities. Both lay eggs (or laid them, in the case of dinosaurs), feathers are an evolutionary development of scales, and chickens have scales on their legs.

Take a look at even the most benign hen and there's something elemental and once removed about her. Chickens are good at hard, unnerving, unblinking reptilian stares. Compare one to a dinosaur with wings, such as a pterodactyl, and you can see that the step between cold blooded and scary and warm blooded and a bit skittish isn't such a huge one.

As time passed various evolutionary tweaks saw new species of mammals and birds appear and in due course dinosaurs became extinct. It's unclear why today's chicken forebears stuck around to survive that particular apocalypse. You might posit that it was because there were fewer huge predators around to think of a chicken as a tasty amuse-bouche.

LATIN NAMES AHEAD

Chickens are part of a bird family known as '*Gallus gallus domesticus*', although one online dictionary consulted during this book's exhaustive research claimed '*Gallus*' was an elasticated brace that held up a man's trousers. Nobody knows if this is true, but if you're the sort of bluffer who prefers their information to be tangential but vaguely credible, stick this in the back of your mental filing cabinet for future use.

Generally chickens are thought to be descendents of *gallus gallus* – a name that fairly trips off the tongue and is more generally associated with wild jungle fowl (but without the '*domesticus*' suffix, for obvious reasons).

Specialist author and chicken keeper Andy Cawthray, who is a genuine expert rather than a bluffer, is worth quoting here: 'The chicken began its existence as a rather localized, obscure, ground-dwelling pheasant in the jungles of South East Asia.'

He suggests that the bird's taxonomy, meaning the scientific classification of organisms and species (there's something to impress your marvelling much-bluffed friends) comes under the heading of Galliformes, which can be applied to game birds in general, and is apparently a group that includes pea and guinea fowl, turkeys, partridges, pheasants and quail. What these birds have in common is that they're foragers who spend most of their time on the ground and either can't fly, or aren't great at getting off the ground for any length of time (although many will roost in trees).

Charles Darwin reckoned *Gallus gallus* was the genetic wellspring from which modern hens sprang, and was the first bird to be domesticated.

IN THE JUNGLE

This applies to their jungle fowl antecedents who, in the case of the red jungle fowl, look and sound very much like the sort of fancy chickens you see looking suspiciously clean and shiny at county shows, although they are less closely related than their looks suggest. Indeed Charles Darwin reckoned *Gallus gallus* was the genetic wellspring from which modern hens sprang, and was the first bird to be domesticated.

There are in fact four recognised types of jungle fowl. Along with the more ubiquitous red jungle fowl species in the genus there's *Gallus varius* (aka 'green jungle fowl'), *Gallus sonneratii* ('grey jungle fowl') and *'Gallus yafayetii'* (Ceylonese, or more accurately, Sri Lankan, jungle fowl, although any non-linguists might think 'yafayetii' sounds Welsh, or even Egyptian, or perhaps Himalayan). Some geneticists have suggested that these birds are also directly related to the hens we see pottering about in urban yards and farmyards rather than just the red jungle fowl, which could well be true, but presumably nobody actually, definitively, knows one way or the other.

Find one of these birds in its natural habitat (particularly red jungle fowl, which aren't actually red, surprisingly enough), and they will look and sound pretty familiar. Domestic chicken-style clucks and assorted avian mutterings will all be present and correct. They will also forage and scratch about for grubs, seeds and other edible things in an entirely chicken-like way. They operate in flocks headed by a cockerel. These will have a pecking order of bird seniority, which will be much the same in their rain forest home as a bunch of Buff Orpingtons in, well, Orpington in Kent.

When danger is perceived everyone shuts up and listens before creating a cacophony of melodramatic vocal hysterics. Females with mothering instincts will go broody to heat up eggs for three weeks to produce the next generation of *Gallus*, etc.

Many geneticists suggest that birds don't have wombs because lugging babies about slows them down. Depending on the breed, domestic chickens can lay hundreds of eggs, but jungle fowl will probably only heave out five or six a year, all of which will be fertile because there isn't much point expending energy on something that isn't useful.

BASE NARCISSISM

Both wild and domestic fowl like to make an effort with their appearance and are helped in this regard by something called the 'preen gland'. This interesting part of the anatomy is, for reasons of convenience, sited near its owner's anus. It secretes an oil that chickens use when preening their feathers. The result is a resplendent feathery coiffure with a sheen that would make any 1960s Brighton rocker proud.

> Both wild and domestic fowl like to make an effort with their appearance and are helped in this regard by something called the 'preen gland'. This interesting part of the anatomy is, for reasons of convenience, sited near its owner's anus.

Another common behaviour is dust bathing. Both wildest and mildest of birds enjoy nothing more than digging a hole, lying in it and thrashing about for a bit. If you're a chicken, the dust might take the sheen from your freshly preened plumage, but scouring your skin with it is an excellent way of murdering the ticks, parasites and mites that plague these birds if given the chance. Sunbathing, where animals stretch out their wings like parasols as they bask in a sideways position is also shared behaviour, and is an effective way of dissipating heat.

There are a couple of other connections between wild and domesticated birds which bluffers need to be familiar with.

UPWARDLY MOBILE

Many, although not all, domestic chickens like to roost off the ground when it gets dark, which is why henhouses have perches. At the end of the day jungle fowl will make for the nearest tree and roost in its branches, which makes them less vulnerable and reduces the likelihood of becoming a passing predator's snack.

Their plumage, and its patterns and colours often make for good camouflage when the bird has secreted itself within a tree. It's amazing how a flamboyant jungle fowl male will be anything but obvious when he's surrounded by leaves and branches – especially if they're moving.

This habit has led to one genetic hand-me-down which if you're a bluffer with a cooperative chicken to hand you can demonstrate to your friends and admirers.

Pick the bird up and move its body up and down and from side to side. Some chickens find this rather alarming and will go rigid. But a reasonably relaxed animal will allow its body to sway and waggle around while its head stays stock still.

In the wild this party trick is very useful for keeping a watchful eye for passing predators. If the bird is perched on a moving branch and most of its body is going with the natural rhythm then it's blending in with the rest of the tree. Having a head that stays still means it's got a better chance of keeping the neighbourhood under observation. Anyone with free range hens will know that when one decides they would prefer to spend the night out rather than being safely shut up in their henhouse it is almost impossible to find them if they're hiding out – especially after dark.

Even chickens who've spent their entire lives in battery cages have this facility, and bluffers can sagely point out that hundreds of years of domestication, interbreeding and general genetic mucking about hasn't eradicated this clever survival mechanism.

IN A FLAP

Another significant shared link between the different birds is the 'not being very good at flying' thing. Jungle fowl and many chickens can get off the ground and remain airborne for upwards of 50ft.

When running along the ground the extra impetus of frenetically flapping wings is the domestic and wild fowl equivalent of a car being fitted with a turbocharger, which provides an immediate and additional boost that makes these birds a lot quicker than they look.

When being chased by something that wants to eat you this is a very useful skill to have. Anyone who has attempted to catch a chicken which is not too keen on incarceration or consumption will know all too well just how rapidly these often ungainly looking animals can travel if they want to keep out of your way.

Which brings bluffers to the question: 'When did human beings first decide chickens were worth catching, and why?'

As far as is known there aren't hieroglyphics involving chickens, although some depicting plucked and decapitated birds of uncertain lineage being offered for sacrifice to the gods do exist, and they look distinctly like oven-ready chickens.

DOMESTICATION

When former US Secretary of State Donald Rumsfeld talked about 'known unknowns', he might have been using his particular brand of linguistic gibberish to shed some light on the animals that morphed into chickens and were first domesticated.

The thing is, nobody is entirely sure when this happened. The odd thousand years is small potatoes in terms of evolutionary history, but those who are slightly more in the know think man and chicken began their professional partnership around 3,000 to 4,000BC.

This is rather later in the day than might be expected. As chicken expert Andy Cawthray points out, animals ranging from dogs to cows were already in the human community by this time. The original reason for the two species' interaction wasn't because chickens are edible, it was actually because male birds engaged in nasty and sometimes fatal punch-ups, and human audiences thought that watching cock birds eviscerating and mutilating each other was fun. Depressingly, some people still do.

In fact, it's only a little over 200 years since Britains embraced chicken farming instead of chicken sparring.

These brutal conflicts had been part of the ancient world's idea of amusing entertainment for centuries. Given that the ancient Romans were keen on feeding Christians to lions and fights to the death involving gladiators, this isn't a surprise and perhaps explains why otherwise perfectly normal people enjoy watching Lord Alan Sugar reducing sharp-suited young hopeful entrepreneurs to weeping wrecks on TV's *The Apprentice*.

DO YOU WANT SOME THEN?

With cock fighting there were numerous gradations of birds, with a particularly demented breed called the Asil being solely the preserve of Far Eastern royalty. There were other breeds with violent tendencies that commoners could own, and these were the preferred fighting bird of those with power and status.

However, as domestic fowls began their march to worldwide domination, human beings began breeding them for show as well as fighting. In Japan ornamental birds have been bred for hundreds of years.

Incidentally, useful bluffer fact, it's thought that the word 'Bantam' has its origins in the Javanese term 'ban ton', which means 'small fowl'.

WALK LIKE AN EGYPTIAN CHICKEN

The ancient Egyptians were probably among the first nations to have kept chickens en masse. During the pharaonic period, spanning over 3,000 years (mostly BC), they were living in incubators, which probably weren't the most congenial environments for either their human or avian occupants. As far as is known there aren't hieroglyphics involving chickens, although some depicting plucked and decapitated birds of uncertain lineage being offered for sacrifice to the gods do exist, and they look distinctly like oven-ready chickens.

FELLOW TRAVELLERS

Violence helped chickens spread around the world as different empires made their way across the planet to rule and conquer. Some of the spoils of war included chickens.

By 700 BC the Persians had diverted themselves by invading India, and returned home with booty including *Gallus gallus domesticus*. By 500 BC Alexander the Great had brought the birds to Europe, where presumably cockfighting had a similar entertainment value then to reality TV now.

The Greeks were so taken by this activity that once chickens were established as a spoil of war, cockfighting was made an Olympic sport. Whether there were any performance enhancing chicken feed scandals history does not recall.*

* Much as the ancient Greek Olympic connection would be pleasing to advance as a little known bluffer's fact, in truth it is almost certainly apocryphal. Bluffers are invited to wryly point this out, using it as further evidence of their extensive knowledge of the world of chickens.

IN THE KNOW

One indisputable moment in world history in which chickens played a central role, involved a decision about military and political policy in ancient Rome. Some chickens were considered sacred and were looked after by priests, or 'augurs'. They would feed the birds and divine if the gods thought battles or policies should be enacted by their response to the food. If they fell on it with enthusiasm all was good; if not, the outlook was dodgy. These birds often travelled with Roman garrisons or at sea.

The Carthaginians trashed most of Publius Claudius's fleet (upwards of 93 boats and as many as 20,000 men were lost), and he returned to Rome for a further kick up the toga and a charge of impiety and blithering incompetence.

During the first of the Punic wars in 249 BC Roman naval commander Publius Claudius Pulcher allegedly jettisoned his unfortunate avian deities overboard, when they didn't come up with the answer he wanted about mounting a surprise attack on the Carthaginian fleet. The birds refused their food. 'Let them drink, since they won't eat,' Publius Claudius is alleged to have said sourly as the unlucky holy fowl were summarily cast into the deep. This ungrateful act wasn't a good move, as the Carthaginians trashed most of Publius Claudius's fleet (upwards of 93 boats and as many as 20,000 men were lost), and he returned to Rome for a further kick up the toga and a charge of impiety and blithering incompetence. What happened next is not entirely clear but either resulted in Publius Claudius going into exile or being let off, but either way he died shortly afterwards. So those birds clearly knew something he didn't or (and this is the most likely explanation) they were suffering from seasickness and so were off their food.

Today we have political psephology and fake news to aid making big decisions and octopuses who are said to be able to work out which teams will do well in the football World Cup. This is otherwise known as progress.

OVER HERE AND OVER THERE

In Britain the Romans introduced very straight roads, running water, under-floor heating and chicken thuggery when the birds conquered the British Isles along with their empire-building owners. When Julius Caesar got off the ferry in around 55 BC, cockfighting became a feature of English life.

Roman soldiers were keen on the sort of heraldry which indicated that they weren't to be messed with. Some legions favoured cockerel motifs to prove to the world just how hard they were. To this day, the Gallic Rooster is one of the national emblems of France. Also known as the 'Coq Gaulois' the truculent chicken decorated French flags during the Revolution, and on Napoleon's standards and army livery as the diminutive Corsican set out to conquer the world. Bluffers will have noted that the Latin word *'gallus'* means chicken (or rooster) and also 'Gaul', the name given by the Romans to the Celtic region of France. Throughout history the humble chicken has traditionally represented an animal that is not to be trifled with.

None of this veneration and historical significance stopped ancient Romans from eating chickens and their eggs, and according to the Smithsonian Institute we can thank them for the practice of stuffing chickens and for inventing the omelette.

It says something about cultural imperatives – although who knows exactly what – that Sir Walter Raleigh brought spuds and tobacco back from America, but when Christopher Columbus found the place one of the things he bequeathed to the Land of the Free, apart from smallpox and possibly influenza, was the chicken.

When not being wiped from the face of the Earth by successive waves of invaders and colonisers, indigenous American Indians took to chicken keeping in a big way, creating new varieties of bird better suited to the local climate and conditions. This was a relatively early example of something that has been a feature of chicken keeping for generations.

SACRIFICIAL

A more recent attempt by aid workers to introduce meatier Western-style chickens, in the shape of the robust Rhode Island Reds (see page 66), to a South American tribe, came unstuck thanks to a long-standing custom of sacrificing the local sinewy birds as a means of divining the future.

The rather gruesome practice involved cutting a chicken's throat and watching how it fell over. A left or right collapse meant a positive answer to whatever question was being posed, but falling forward meant a negative one. The hefty Rhode Island hens always pitched forward, so the locals stuck to indigenous breeds.

This sacrificial practice was recorded by the Smithsonian Institute, the world's largest museum and research centre, along with a 1993 Floridian court case involving practitioners of the Santeria religion, which is an amalgam of Catholicism, Caric and West African Yoruba faiths, and takes in chicken sacrifice (along with other unfortunate animals including turtles and guinea pigs).

A local order in the town of Hialeah banning animal sacrifice was overturned after a local Santeria priest fought the ban, which resulted in predictable argy bargy between civil-rights organisations and animal lovers.

'Religious beliefs need not be acceptable, logical, consistent or comprehensible to others in order to merit First Amendment protection,' ruled the presiding justice, which was good news for practitioners of animal sacrifice, but bad news for chickens living in the Sunshine State.

FASHION STATEMENTS

By the mid-1800s when Charles Darwin's theories of evolution were creating a bit of a stir, people started thinking about animal genetics, and how domesticated birds could be improved, or changed.

Queen Victoria had luxury cast-iron henhouses built in a Gothic style for Osborne House, her country retreat on the Isle of Wight. It is safe to assume that these were largely predator-proof, unless the door was left open or the foxes were carrying metal cutters.

Chickens were not excluded from this rush to eugenics, which in Victorian Britain manifested itself in breeding designer birds that looked pretty or interesting. This was known as the 'fancy fowl movement'.

A great many birds that people covet today made their debuts around this time (think Buff Orpingtons and Brahmas with their paisley-like feather patterns). This was also the era when the Poultry Club of Great Britain was founded. By 1865 this august and still very extant body had created what it described as 'a standard of excellence'. This measure of chicken body narcissism became the benchmark for what was considered ideal characteristics of the many and varied fancy fowl that were being hatched by men with large whiskers and top hats.

One keen fancy fowl keeper was a certain Queen Victoria, although presumably she had minions to muck them out. She also had luxury cast-iron henhouses built in a Gothic style for Osborne House, her country retreat on the Isle of Wight. It is safe to assume that these were largely predator-proof, unless the door was left open or the foxes were carrying metal cutters.

TAKING THE STRAIN

By the time people saw chickens as a source of food rather than blood lust and betting, fans of chickens in Italy and Britain were hard at work creating breeds that were 'farm friendly'. By the mid-1800s their American counterparts were doing much the same – and this can be identified as the era when many of the working breeds that led to the chickens knocking about today began to emerge. Such animals carried more flesh and laid more eggs.

Industrialised conflict, in the shape of the First World War of 1914–18, brought with it the need for similarly industrialised farming and a rapid increase in food production to feed massive, mechanised armies. For farm chickens, the lay more eggs/get fatter faster imperative was well and truly in place by this time, and American farmers and geneticists in particular interbred birds with these characteristics. The Rhode Island Red is one of the results of their efforts, although their uselessness as sacrificial diviners of the future wasn't fully known at the time.

WAR BIRDS

After a 21-year break European hostilities re-commenced when the Second World War broke out, by which stage factory farming in a form that would to some extent be recognisable today was in place.

This is a trend that has continued ever since. From the interwar years until well into the middle of the 20th century, technology drove the trend towards factory farming where food was essentially mass-produced along with almost everything else – from cars to canned fruit. Pesticides, vitamins and antibiotics have all played their part in making it possible to keep more animals in close proximity without them getting ill (or at least less of them getting ill). The lessons from the wartime shortages and rationing showed that while townies were more than willing to keep hens and rabbits for eggs, meat and fur, they frequently hadn't the heart to kill them. This was a part of the war effort where many otherwise dutiful and stoical Brits were found wanting. But the new pets certainly weren't complaining.

BATTERY JUMP-START

Keeping a lot of chickens together in barns meant that their more anti-social habits, which included bullying and cannibalism, were harder to control and monitor. Putting birds in stacked, connected cages meant that farmers could keep an eye on them. In practical terms there were hygiene and egg collection advantages and recognisable battery systems were being used in the USA by the early 1930s. Today, many have lighting systems so that they are illuminated at night. Chickens do not lay eggs in the dark and this increases the number of eggs produced. Rather disturbingly, similar systems have been used for farmed animals ranging from mink to rabbits.

Since 2012 in Europe chicken battery cage sizes have grown a little bigger than the A4 sheet of paper per bird floor area (roughly 12 by 8 inches) previously permitted. Known as 'enriched' cages, these have some perching and dust bathing areas, but the debate about the ethicality of these systems continues. Many people passionately believe that keeping social, foraging animals in confined, sterile conditions is cruel.* There is also a debate about the health implications for both egg laying and broiler birds (those reared for the table) with concerns about stress, heart and lung problems and brittle bone disorders caused by lack of calcium. Proponents point to the hygiene, husbandry and efficiency aspects of these systems, and the fact that factory farming has undoubtedly made food cheaper. Chicken was once a wealthy person's food, but no longer.

EGGS AND ETHICS

All this has taken bluffers into rather serious territory for an apparently jolly and entertaining book about chickens, but these things are all part of the art of

* For a fuller exposition of the arguments for and against factory farming see *The Bluffer's Guide to Veganism*.

keeping them. How many of these birds are kept in factory farm conditions? You will find figures of between 300m and 500m worldwide, with some countries, such as Argentina and the USA, using cage systems for the vast majority of commercial egg sales.

In Britain according to the Compassion in World Farming campaign the number of battery chicken eggs bought has fallen from just under 80 per cent in 1998 to around 50 per cent by 2013. Two years before that the abomination known as beak trimming (done to reduce feather pecking) was made illegal in the UK although other countries, such as the US, still permit it.

For UK chicken bluffers who want to major in agriculture, 1947 is an important date. This was the year Britain introduced the Agriculture Act, which gave farmers subsidies for embracing new technology that reduced British dependence on imported meat. Being an island nation that had recently survived six years of total war that included sustained attempts to choke off Britain's supply of imported food, this was not a great surprise, but this was one of the main drivers of high-yield farming systems, including chicken batteries.

There's an opportunity here for bluffers to point out that the country already had a perfectly sustainable and inexpensive means of producing a highly nutritious organically reared foodstuff, not unlike chicken to taste, which had served the country well during the war. It was called 'rabbit', a heroically procreative long-eared, short-tailed, burrowing mammal of the family *Leporidae*. Unfortunately it had an irksome habit of eating farmers' crops, and so it was considered to be a pest. Coincidentally (or not) it was all but wiped out in the 1950s by a viral disease called myxomatosis which effectively decimated the wild rabbit population and had horrifying side effects among those which weren't fortunate enough to die within a few days after infection. Not all farmers were hugely sympathetic about the rabbit population's plight which, curiously, seemed to afflict rabbits only. Bluffers are allowed to conjecture about whether the dark arts of Porton Down's biological research facility had been employed in ensuring that the disease didn't jump species and only infected rabbits. Bluffers can also speculate about what might have happened if rabbits on the UK mainland had survived relatively unscathed: Kentucky Fried Bunny? Bunny Nuggets? Rabbit Tikka Masala?

CONSPICUOUS CONSUMPTION

Moving on from ailing bunnies, any chicken bluffer worth their egg timer will know that nations with the biggest populations aren't necessarily the ones that have the biggest appetite for chicken eggs. A visit to the information cornucopia otherwise known as the World Atlas website will reveal that the nations whose populations are most likely to be egg lovers are spread all over the planet, and some of them are quite small. This is vital information for bluffers, because so-called experts in the art of chicken maintenance are unlikely to have a global view of egg consumption. In fact they're unlikely to have a view beyond the end of their garden. So the top per capita egg consumer isn't China, but its stripling geopolitical rival Japan. According to World Atlas, the average Japanese citizen gets through 320 eggs a year, and with a population of 120m that's a lot of eggs and a lot of chickens, which is why Japan's human/hen population is estimated to be about 50/50. Some of these eggs are enriched with minerals, and most, apparently, are downed raw like oysters.

The next stop on the egg-eating bluffer's world tour is Paraguay, the second biggest consumer per capita. The average Paraguayan citizen will get through 309 eggs a year. That's still nine more than the average Chinese citizen consumes annually. Although technology and urbanisation seem to be Chinese watchwords these days, many of those eggs will have come from small, rural farmsteads.

For rapid-fire egg-related bluffing, the rest of the statisticians' national egg eating top ten is made up of Mexico, Ukraine, Malaysia, Brunei, Slovakia, Belarus and the Russian Federation. There could well be a trading partnership about to emerge from this list of nations bound together by their love of eggs.

The UK's farmed chicken egg production quantities barely register as a statistical pimple on worldwide figures, but the British Egg Information Service ('the official voice of the British egg industry'), estimated that in 2018 Britain's 41million working chickens helped its 66m-strong population eat an annual average of 199 eggs per person (up from 171 per egg-eating punter 12 years before). That means Brits got through 13.1bn eggs annually and 36 million every day.

British chickens laid some 87 per cent of them, and

63 per cent of British retail sales were for eggs from, allegedly or actively, free-range birds – that's around a 50 per cent rise in 14 years. And this, of course, doesn't account for the many domestic urban chickens where eggs are not bought and sold in any conventional currency (unless you count barter transactions which, to a large extent, underpin the trade in back garden egg production).

Britain imports 1,897 million eggs a year and exports 188 million. Its domestic egg market was worth £1.035 million over the same period, so it's no bluff to say that the unassuming chicken is a significant earner.

The Poultry Club of Great Britain maintains that true bantams have no equivalent full-sized counterparts, so presumably those that do are false bantams, and these cherished birds are bred to be looked at rather than eaten (although they could be in an emergency).

MEET THE FLOCKERS

There are hundreds of different chicken breeds, from variations of hybrid farm hens that are either expert at producing a lot of eggs or meat. There are bantams, heavy, and fancy chicken breeds, and a committed chicken bluffer will have a head full of facts about them, but to have those facts, it's important to know a bit about individual breeds.

The Poultry Club of Great Britain reckons that there are 'well over' 100 varieties of chicken in the UK, and that they're bred as large fowl and bantams. The club maintains that true bantams have no equivalent full-sized counterparts, so presumably those that do are false bantams, and these cherished birds are bred to be looked at rather than eaten (although they could be in an emergency). Chickens can be classified as being hard and soft feather breeds, with the former descended from game bird stock, so they have a disreputable family history of cockfighting in their family trees. Apparently those birds with respectable and benign family trees have soft feathers.

This guide has compiled a compendium of chicken

types, which might be described as an A to Z of avian oddity. If you're a true chicken expert rather than a bluffer, you will spot some gaps, but this is because a full list of breed types would take up the entire book, the purpose of which would be to dazzle readers with even more unmissable but varied hen information. In this instance, however, less is definitely more.

Asil. One of the oldest domestic fowl breeds, there are records of these birds being bred for cockfighting in India 2,000 years ago.

The cockerels have elegant plumage, a powerful build and can be extremely bad tempered with each other, to the extent that when things kick off, fighting can last for days at a time.

This behaviour (not unlike that exhibited by the knight from *Monty Python and the Holy Grail* who kept fighting as his opponent lopped off various limbs) is, thank goodness, no longer prized by most owners, many of whom keep these elegant animals as show birds (separately from other Asils).

Amber Lee. This bird could be named after a member of a girl band, but is in fact a hybrid chicken capable of laying an exhausting 330 plus eggs in its first year, although after such an effort they can be excused if their productivity drops off in subsequent years. In character, the Amber Lee is sociable and nosey.

Ancona. These feisty birds are named after an Italian city and first appeared in Britain during the 1850s when keeping fancy fowl became very fashionable. Apparently they made their British debut at the Great Exhibition of 1851. Anconas come in regular and bantam sizes, and are very keen on foraging, a fact that potential chicken keepers who are keen on keeping their gardens unlike a bomb site might consider. They can lay up to 270 eggs a year and are instinctively showy, so both food fans and people who exhibit birds like them. According to Poultrykeeper.com the Ancona's large comb is also susceptible to frostbite.

Andalusian. If you want to show off your knowledge of rare chickens the Andalusian is good for a name check. As with the Ancona the bird is named after a place, but this one's a vast region in the south-west of Spain. The bird arrived in Britain in about 1847 and is not a prolific layer, although it will gradually get going producing eggs from an early age. It has a long body and a game bird's gait, and rather pretty, interlaced plumage thanks to years of deliberate interbreeding. Generally not prone to hysterics this is a bird that will be good at giving its owners the slip because it's very fast on its feet.

Appenzeller Spitzhauben. Although it has a name that sounds like a German mouthwash, the Appenzeller Spizhauben is actually a Swiss chicken that only appeared in Britain in 1982. These birds have distinctive head feathers and are apparently named after a Swiss lady's bonnet that goes with traditional dress. The gentlemen Appenzeller Spitzhauben chickens have the additional distinguishing feature in the headgear department of a double-spiked comb. It looks nothing like a traditional Alpine hat.

Araucana. A bird with Chilean and Scottish ancestry, the Araucana is a loner and can be quite standoffish. Like the Andalusian it can also be a fast mover, so catching one can be time consuming. In the 1930s it was bred to produce bluish-green eggs that are unique in that the colouration goes right through the shell. Other coloured hens' eggs have white inner layers apparently. Speaking of laying, free-range Araucanas like to secrete their eggs in secret stashes, and are therefore not always easy to find.

Some Aruacanas (and indeed a number of other breeds) are known as 'Rumpless,' not because they are lacking in rumps, but because they have no tail feathers. This sawn-off look is the result of selective breeding rather than a nasty accident involving a farm implement.

Ausaberger. A black chicken that first appeared in the Augsburg region of Black Forest in Germany in about 1870. It has a face as red as someone who writes in to complain to the BBC about people on Radio 4 starting every sentence with the word 'So', and has a curious double comb. Bred to show or be eaten, you don't see many in Britain.

Australorp. This New World bird's country of origin is (not so subtly) hidden in its name. Bred from very English Black Orpington stock, but a bit smaller, Australorps have developed a serial egg-laying habit. An Australorp reportedly holds Australia's egg-laying record, having produced 364 eggs in 365 days. Unsurprisingly bulging eyes are another feature of the breed.

Barnevelder. A Dutch chicken that was first hatched shortly before the First World War got under way, it reached Britain at the start of the 1920s. These birds are productive layers, and generally have laid-back personalities, despite being able to fix you with a baleful stare from prominent, orange eyes.

Black Star. This hybrid bird has a reputation for physical robustness and a disease-resistant immune system. Some are black, unsurprisingly, but there are feather variations taking in greens and copper feathers around the chest and neck. They are bred from Rhode Island or New Hampshire roosters and a Barred Rock hen.

Bluebell. Another hybrid created from encouraging specific breeds of chickens to get it together. In this case the Rhode Island Red and the Marans. The result is a stolid animal with feathers that come in a variety of grey tints. Probably thanks to the Rhode Island Red side of its lineage this bird will lay upwards of 250 eggs a year. Unfriendly Bluebells are unusual.

Bovans Goldline. Any self-respecting chicken bluffer will identify the Rhode Island Red as having genes that have played an important part in creating today's working farm chickens, and these birds are related to other breeds that were bred en route. This curiously named chicken is a good example, being the progeny of Rhode Island and Light Sussex stock. These birds are physically tough, friendly, and in their first year the girls can pop out over 330 eggs.

Brahma. Despite being given a name that owes its provenance to the Indian subcontinent, these big, lumbering, generally gentle birds actually originated in America. They made their ponderous way to Britain over the Atlantic – presumably by ship – arriving around 1852. These easy-going heavyweights make ideal pets because they're pretty friendly, but laying lots of eggs isn't a Brahma strongpoint, although the females are keen on parenthood, so broodiness is not uncommon.

Buttercup. Small, flashy looking and of Italian extraction, these birds don't lay a lot of eggs and don't go broody, so the strain has been kept going by people stuffing Buttercup eggs under foster chickens of different breeds or using incubators.

They look very pretty with colourful, intricate plumages, and are prized as ornamental fowl. They like being free range but are less than keen on people so tend to make themselves scarce, which is a major demerit on the ornamental front. In winter their combs can suffer from frostbite.

Cochin. A Chinese émigré that reached the US and Britain during the 1840s and was originally known as Chinese Shanghai Fowl. These birds are big and imposing with an abundance of colourful, showy plumage, giving both hens and cockerels a rather exotic appearance.

This riot of feathers extends to their feet, which are covered in feathery flares which can get damaged in dense undergrowth, so Cochins like nothing better than a lawn to sashay about on. They don't lay a great many eggs – around 120 a year in their prime – so take longer to peg out than some breeds, living up to ten years.

Cornish/Indian Game. Bred for the table, these powerfully built creatures have particularly grumpy expressions thanks to sunken eyes and a down-in-the-mouth beak shape.

Having a relatively thin coat of feathers they tend to be grumpy during hard winters and will feel the cold more than some chicken breeds. Otherwise they have engaging personalities.

For obvious reasons these hefty creatures need low perches and benefit from henhouses with decent-sized door openings to prevent 'wedged in the hole' indignity. They don't lay that many eggs but will go broody.

These birds are sufficiently broad beamed and short necked so have trouble preening their derriere feathers, which means mites around the backside can be an issue. This sort of insider knowledge might be faintly revolting, but it is the very essence of avian bluffery.

Croad Langshan. Major F.T. Croad wasn't a man burdened with false modesty, so when he brought this breed of chicken back from the Langshan region of northern China he named it after himself. This was in 1872 when the birds had black plumage with a hint of green. Now, thanks to breeders, there are white varieties

– all sharing a relaxed demeanour, which might explain why they're reputedly Australia's most popular chicken. They need a lot of space but oddly enough, despite their runway-sized enclosures, they can't fly.

Dorking. This stolid, placid bird has the misfortune to be very good to eat, and many of them were bred for the table. But their genetic inheritance goes back a long way, as their forebears arrived in Britain with the Romans, when presumably Dorking was a few wattle and daub huts in a wood, rather than a traffic-jam-prone commuter town in Surrey next to the A24. Generally Dorkings have a relaxed approach to life, which makes them good pets, but they do like scratching about and need plenty of space.

As with a number of 'heavy breed' chickens, Dorkings have an ambivalent approach to egg laying, producing about 140 a year.

Dutch Bantam. These diminutive birds have a habit of bonding with their owners, so anyone with limited space and a desire to make friends with some chickens could do a lot worse than consider the Dutch Bantam.

Anyone wanting freshly laid eggs in the winter should look elsewhere, as the birds only lay in the summer. Another quirk is the 20 days' chick gestation period inside Dutch Bantam eggs. Most chickens take exactly three weeks.

Faverolle. Originating from France these hefty animals have feathery beards and good social skills so that they often bond with their owners, which makes them especially child friendly. With little wings, short necks and barrel chests these chickens have the stance of nightclub bouncers, but aren't great at flying, which makes them an easier prospect to keep in check than some flightier birds. A mixture of Dorking, Asiatic and Houdans stock, these birds are named after the French village where they were originally bred as farmyard chickens.

Fayoumi. When not erecting huge pyramids the ancient Egyptians were raising chickens, including the Fayoumi, which has been around for a very long time, although it only reached Britain in 1984.

If caught the breed is well known for emiting blood-curdling screams, but catching them in the first place is a bit of a challenge because these birds are particularly adept at flying and can be very effective escapologists.

Frizzle. This appropriately named bird has feathers that curl backwards, giving it a puffed-up appearance as if it's survived a washing machine's spin cycle and then been blasted with a hair dryer.

These characterful animals are popular at chicken shows, where when competition is involved, birds with the curliest feathers win the prizes. Apparently there are recognised feather types, which is worth knowing for bluffing purposes. These are 'frizzled', 'over frizzled' and 'flat coated', but bluffers will observe that only frizzled Frizzles can make the grade in chicken exhibitions.

For lay chicken keepers, Frizzles are at the comic end of hen physiognomy. Perhaps this is one reason why the breed is now comparatively rare.

German Langshan. Back again with Major Croad and the Langshan chickens he brought to Blighty from northern China in 1872. Seven years after that these birds arrived in Germany where they were crossed with black Minorca hens to encourage more egg laying, which was relatively successful as the females will knock out about 180 eggs a year. The cockerels have extravagant tail feathers and the breed is known for its hardiness.

Gingernut Ranger. If there was a contest for daft chicken breed names the Gingernut Ranger could well win it with a moniker that might be applied to a minor character in a Mel Brooks cowboy movie ('The Gingernut Ranger vs the Eccles Cake Kid'?). Some chickens have five toes, but the Gingernut Ranger seems unconcerned about only having four on each foot. When tamed, these birds are happy to follow their owners about.

Hamburg. Despite its name the Hamburg actually originates from Holland and has been around for a long time.

The Hamburg's present-day shape and plumage combinations can largely be attributed to British fancy fowl breeders who interbred different strains. There were regional variations, so Hamburgs in Yorkshire and Lancashire have different plumage (and accents) from each other and their cousins in the rest of Europe.

Houdan. Another chicken named after a place, this time in north central France. This breed has been around since the 1800s and was bred for markets in

Paris and London. The Houdan looks pretty similar to the Dorking, and could possibly be directly related as both have a fifth toe. Unlike some chickens that have a fit of the screaming abdabs if picked up, Houdans enjoy a fuss – especially if they have been hand reared and are used to people.

Ixworth. This bird was developed in 1932 by one Reginald Appleyard who named his feathery creation after the Suffolk village where he lived. To produce the Ixworth he persuaded White Sussex, Orpington and Minorca chickens to get it together, along with Jubilee, Indian and white game birds. The result is a notably unflappable bird, productive in the egg department, good to eat and these days pretty rare. A bantam version has all but vanished.

Jersey Giant. Officially the world's biggest chicken breed, this epic bird doesn't come from the Channel island that gave the world the 1980s television detective *Bergerac*. It originates from New Jersey in the USA. Apparently these chickens were known as Jersey Black Giants (which sounds like an American football team) and first appeared in 1870. They are the result of deliberate interbreeding between Brahmas, Javas and Black Langshan chickens. The word 'exotic' springs to mind.

La Fleche. Hailing from the French valley of La Sarthe, this bird is a breed engineered for meat and egg production. It's keen on flying and can rise around 2m into the air, so will require open prison-style fencing if it's not going to start land-grabbing where it isn't wanted. You can feed, accommodate and generally fuss these birds, but it won't do you much good because they aren't keen on people. They lay around 200 eggs a year by way of compensation.

Lakenvelder. This chicken has been trundling around in Germany since the 1830s. Lakenvelders tend to regard human beings with suspicion, keep to themselves and are good at flying over fences so wing clipping or prison camp-like fences of 2.5m plus will be needed to keep them where you want them. These birds are less placid than many breeds but are hardy and robust.

Legbar. These birds were first bred in the 1930s, and their big, upright tail feathers, floppy combs and bluish eggs point to their being related to Araucanas with Brown Leghorn and Barred Rock DNA stirred into the mix. This is an 'autosexing' breed, which does not suggest a form of avian onanism, but means that when hatched the colour of their chicks' down gives away whether the baby birds are male or female.

Leghorn. Originally from Italy, but with English, American and Danish variations, the Leghorn is a very vocal chicken. They're good at flying, and perhaps nostalgic for their wild past will sometimes roost in trees. These birds are enthusiastic egg layers but are less keen on parenthood, as the females rarely go broody. Foghorn Leghorn, the beefy rooster in *Looney Tunes*, is probably the most celebrated member of the family.

Maran. Another bird that originated in France, and once again named after a town, Marans have made their home in Britain since the late 1920s.

Most UK Marans have 'cuckoo'-coloured grey and silver plumage and they lay distinctive, dark brown eggs. These are feisty, independent birds and it's not unheard of for them to have rows with other hatch mates. Unlike some breeds they are unlikely to bond with their owners.

Marsh Daisy. This chicken originally hails from Southport (or to be strictly accurate for bluffing purposes, Marshside), so it is almost Scouse. Dating back to the 1890s the breed is an amalgam of six different chicken types including the Sicilian Buttercup. These days the Marsh Daisy is so rare that it has been put on the Rare Breeds Survival Trust's endangered species list.

Minorca. These birds possess enormous white ear lobes and extravagant combs and wattles. Originating from the Balearic islands these big powerful chickens used to go by the rather insulting-sounding name of 'Red Faced Black Spanish'. The Minorca is the largest and strongest of the Mediterranean breeds which these days are kept mainly for show but they also happen to be producers of some very large eggs.

Naked Neck. A chicken that few would describe as conventionally beautiful, the Naked Neck lives up to its name and is also known as the Transylvanian, with the obvious vampirical connotations that the description implies. The breed hails from Hungary via Germany where a strain with non-feathered necks was perfected. In fact the Naked Neck has 50 per cent fewer feathers than many other chicken breeds, and is said to exist because it took less time to pluck. Funnily enough they find really cold weather a bit of a trial.

New Hampshire Red. This American bird is a cousin of the better known Rhode Island Red and was developed in 1915 when the First World War meant that farmers and food scientists were keen to speed up the maturity of farm animals to aid food supply; this animal doesn't take long to grow up. Unlike some chickens, who like a bit of a ruck, these birds get on well with each other and people.

Norfolk Grey. This breed had virtually died out by 1974, when a flock of four were discovered some 50 years after they had first been exhibited.

Understandably, given this low population base the Norfolk Grey is a rare chicken now, but rather a pretty one. The cockerels in particular have a showy, 'look at me I'm fabulous' plumage.

Old English Pheasant Fowl. These catchily named animals have a penchant for adventure and will roost in trees and high rafters if given the chance. They gained their current name in the Edwardian era, having previously been known as Yorkshire, Copper Moss, Golden and Old Fashioned Peasant Fowl, depending on which part of the country they were living in. The breed has been around for a long time, and is generally happy around people unless there are chicks in the offing, in which case relations cool.

Orpington. The original Victorian commuter-belt chicken, the Orpington first appeared in 1886. These gentle, well-padded creatures are the result of top hatted chicken fanciers cross breeding Black Langshans, Black Minorcas and Black Plymouth Rocks. Unsurprisingly the first Orpingtons were black.

Further genetic mucking about resulted in the yellowish Buff Orpington in 1894. White varieties followed five years later.

The Orpington's temperament is horizontal with extra relaxation, and the females make good parents. These birds like their grub, and will shovel it down like chocolate Labradors, carrying a fair bit of weight as a result. This means that they need plenty of exercise to avoid joint problems and fitness issues.

Plymouth Rock. This bird was doing its thing in Plymouth, USA, in the late 1820s, and the animals still scratching around today share a lineage which includes breeds whose ancestors still live in Javanese forests.

There are plenty of varieties of Plymouth Rock, with a selection of colours. One branch of the family is known as the Buff Columbian, which sounds like an illegal substance rather than a chicken.

The bird's intricate plumage is one reason for its enduring popularity, with the Barred Plymouth Rock's wiggly patterned black and white feathers which end in black tips – winning it plenty of chicken couture fans.

Rhode Island Red. An important bird on the genetic road to the modern farm 'hybrid' chicken, this glossy-feathered animal was conceived during the First World War to grow up fast, lay plenty of eggs and be portly in an edible way.

Its parents are a multinational bunch, including Asiatic, Leghorn and Red Malay game birds, and the Rhode Island is a dual-purpose bird, which means people breed it either to eat its meat or its eggs.

These birds have a reputation for being physically tough and even tempered. They are the strong and relatively silent types of the chicken world.

Scots Grey. A chicken with a reputation for being a bit of a tough nut, capable of surviving in an inclement climate, this breed dates all the way back to the 16th century. Long legged and rather upright, the Scots Grey's posture hints at game bird ancestry. It likes the great outdoors and will roost in trees if given the chance.

Speckledy. Another hybrid chicken with some Rhode Island Red blood in its veins, the Speckledy is a cross between the Rhode Island and the Maran, whose feather patterning it has appropriated, but not this bird's relative disinterest in laying eggs, which are dark brown and often speckled. The shells are hard so will require more spoon bashing if boiled.

Sumatra. These rather elegant birds started life as jungle dwellers in Sumatra, Java and Borneo.

With their double spurs and athletic build the cockerels look as if they'd be up for a ruck, but in the wild this only happened during the relatively short breeding season. The rest of the time these territorial birds got along pretty well.

Sadly for them local people chose the breeding season to catch the males for cockfighting. Those that survived were released back into the wild when their aggro-promoting hormones had died down.

The birds reached Europe and the US in the late 1840s, where once again people enjoying avian punch-ups were a driving force for their arrival. They were in Britain by 1900 when this sort of brutality was outlawed. These days they are mostly kept for their undeniably good looks.

They're diligent parents, will keep laying eggs when it gets cold, and perhaps as a nod to their jungle past have helicopter like vertical take-off skills when confronted by predators.

Sussex. This is quite possibly the oldest chicken type in Britain, with examples being bred in AD 43 when the Romans were strengthening their grip on the country. These were bred to eat but became dual-purpose eggs and meat chickens and they appear in various shades including buff (or brown), red, white, speckled and silver. Some chicken breeds are permanently up for a fight, not so the Sussex, which is generally a calm, easy-going creature. They like doing that chicken thing of tramping about and foraging, but don't go stir crazy if they're penned up.

Take a Sussex and pair it with a Rhode Island Red and the resultant chicken children will be Sussex Star hybrids, which are the acme of calm and egg-laying prowess. A healthy female should produce an eye-watering 250 eggs every year.

Vorwerk. The bird didn't appear in Britain until the go-getting glam 1980s, but had been doing its thing in German farmyards since the beginning of the last century. They often have neck and head feathers that are a different colour from the rest of the bird, which gives the impression of a chicken cut and shut, with the head of one bird grafted on to another, but they still posses a certain dignity.

Vorwerks can lay 170 eggs a year and people in Germany are happy to eat them. A similar fate is less likely in Britain where the breed is extremely rare.

Welsummer. These birds originated from the Dutch village of Welsum, and first made it to the UK in 1928.

They are feisty, characterful and the females tend to make good mothers. They can be long lived (up to nine years or more), although harder-hearted potential owners might note that they could spend two-thirds of that time not bothering to lay eggs.

Wyandotte. This bird is an American export, named after the Wyandotte American Indian tribe, although there isn't a direct historical connection between the bird and its human namesakes.

There are upwards of twelve variations in plumage design and the birds are popular among small flock owners and more hardcore hen fanciers, some of whom bring their favourite chickens to county shows. The Silver Laced variety is arguably the prettiest and most productive of the breed (the intricate lace effect of the feather pattern is worth seeing). It was the result of organised nights – or more accurately 'days' – of passion between 'silver–laced' Wyandotte hens and crossbred Cochin-Brown Leghorn cockerels.

A good chicken bluffer, when visiting a chicken-keeping newbie, will be able to look at their choice of henhouse and make some salient points that the recipient might find useful, dispiriting, or even faintly irritating. Depending on the mindset of the advice giver, all these are potentially satisfying outcomes.

WHAT YOU NEED

A true chicken bluffer might not actually own any birds, but that won't stop them from knowing what's needed to do so in terms of kit and husbandry.

First consider housing. Generally speaking chicken hutches are rather eclectic in style. They range in design from avian prison cells to deeply twee country cottage/ designer doll's house specials, but a good chicken bluffer, when visiting a chicken-keeping newbie, will be able to look at their choice of henhouse and make some salient points that the recipient might find useful, dispiriting or even faintly irritating. Depending on the mindset of the advice giver, all these are potentially satisfying outcomes.

A good chicken house is secure, easy to keep clean and has plenty of air space, which means that perennially popular chicken arks (the triangular ones that look like giant Toblerones) are at something of a disadvantage.

Chickens can suffer from dodgy lungs (see the chapter on avian ailments). This has to do with how they breathe, but the upshot is that chickens are quite prone to respiratory problems and these can sometimes prove fatal. Where they roost can have a lot to do with this.

ARK ANGST

So, if your birds live in a triangulated henhouse, known as an ark, this can be an issue. The ark's shape means they will roost in the roof space, and the amount of room there means birdy heads are close together and there's limited air space. This creates a breeding ground for spreading lung-related nasties.

Henhouses that have plenty of head space won't eliminate this risk, but as a chicken bluffer you can sagely point out that they will at least reduce it and the birds will, quite literally, have more breathing space, although some will sabotage things by snuggling up together.

Some people use old, or even new garden-style sheds as chicken accommodation. These have the advantage of being big enough for the birds' owners to get inside to clean them out, they are relatively easy to assemble and relatively cheap to buy – certainly compared to henhouses at the garden playhouse end of the market which can cost hundreds. But beware.

Many sheds have chipboard-style flooring. For a chicken-free shed this stuff is perfectly serviceable and will last for years, but it does not respond well to chicken guano, which contains a cocktail of urine and ammonia that will in due course rot the hardiest timber. Something made of compressed soft wood sawdust stands no chance against it, and will rapidly become a mouldering, soggy biomass.

In the end a visiting chicken keeper will put his or her foot through the floor and get a sock full of damp, chicken urine-marinated wood chips. You don't need to be a chicken bluffer to know that this is very irritating.

However, you are probably thinking ahead. Images will be filtering into your mind of putting down straw and sawdust to make the shed a nicer environment. This will certainly mop up some of the muck. Some owners lay down sections of cardboard or even newspaper over henhouse floors, but both materials share compressed timber's capacity to absorb liquid and exude a damp fug. This miasma of fetid air and water droplets is not good for birds with a susceptibility for dodgy lung infections.

A henhouse with exterior grade wood, or even marine-ply grade timber flooring is better news, and having some sort of metal or plastic trays above it is better still. Some bespoke birdhouses have floors that are effectively trays that can be slid out for cleaning purposes.

DOWN AND DIRTY

Laying down straw and sawdust will make any henhouse a nicer place for its residents, but this stuff has to be shifted at least once a week, or it will turn into a nasty, festering pile of ordure much loved by rats and other vermin. As previously mentioned, hay is a non starter for chicken house bedding because it contains spores that can do nasty things to chicken respiratory systems.

Another need-to-know aspect of keeping a chicken house spruce and bug free is regular recourse to a scrubbing brush and suitable, animal friendly disinfectant, often diluted in water, to keep everything clean.

By now your brain's chicken knowledge synapses will be snapping at full megawatts, and you will remember the warning about henhouses made from flimsy wood. This clearly won't respond well to regular dousing with disinfected water.

CRACKING UP

Many wooden henhouses are made from lapped soft wood or tongue and groove. These are carpentry terms and relate to how bits of wood are joined together when making stuff, including chicken housing. Even the chickens know this.

Many sheds have lap wood panels that are partially overlaid. The floorboards in your house are probably connected by tongue and groove joints, where the tongue of one piece of timber slots into a groove cut into the side of the next. There's nothing particularly wrong with these chippy techniques per se, but they are a mixed blessing in the context of chicken accommodation.

The problem is a nasty bug called red mite. Once again the chapter on ailments will give you the lowdown on what these creatures do to your chickens, but they like nothing better than to jump on to the birds and suck their blood.

Given the chance, they will multiply with the speed and determination of wannabe TV talent show entrants and they will willingly do this between overlaid bits of timber and in tongue and groove joints, where they are very hard to reach. In the days when creosote was considered a benign garden staple rather than a dangerous carcinogen, this could be slapped on to red mite infested timber, and hundreds of them would come bubbling out of their hiding places and expire. Creosote has been banned since 2003 and replacement organic pest murdering products will do red mite in eventually, but they often require several applications before the little varmints croak.

Beyond being kept properly clean to start with, the best wooden henhouse will be made of big, simple sections of timber with as few joints as possible. The less of these there are, the fewer areas there will be for red mites to secrete themselves.

This also relates to roofing felt. Anyone making their own chicken hutch might be tempted to cover the whole thing in the stuff to keep out the weather, but this will create an often inaccessible bug breeding ground between felt and woodwork.

None of the above means that the resulting chicken hutch has to be a style-free box, but it will make life simpler for chicken owners and nicer for their birds, which tend not to venture an opinion about their accommodation. So, if you're in bluffing mode and are invited to venture an opinion on a henhouse, bear these facts in mind, and while you're at it, look at the house to see how easy it actually is to clean.

You can point out that once the novelty of something that looks like Disney World's Cinderella castle has worn off, and the owner is suffering from a herniated spinal disc from the contortions needed to reach the roof space with a scrubbing brush, they might well have wished they'd checked out how easy it was to clean before they'd bought it.

PLASTIC FANTASTIC

And then of course there are plastic henhouses. You will be able to point out that these are, generally speaking, far less mite friendly, and will be easier to keep clean, and possibly less hard work to maintain. It doesn't take an avian sage to work out that chicken guano is harder to remove from wood than plastic. Also, plastic houses won't rot and fall to pieces as a result of exposure to the nasty stuff that comes out of the back of a chicken. Vinyl and laminate are also options, particularly for floors.

So far so desirable, but you might reasonably point out whether something relatively light will have the staying power to withstand a puff of wind if it's located in an exposed location.

There are of course different henhouse designs, including heavy-duty jobs made from re-cycled plastics and held together with pins and clips. Some of these will sit it out during a hurricane if they are properly anchored, but as you will have already worked out, that does not make them easily portable and this might be an issue for someone with visions of regularly relocating their hens round what remains of the lawn.

As a chicken bluffer you can say something along the lines of 'one person's solution is another person's problem', and wait to see if the recipient of this homily will reach for a handily placed mallet or not.

IT'S GOOD TO SHARE. NOT

Assuming that a henhouse move is not an issue, it might very well not go anywhere at all, which for free-range chickens is perfectly fine. They will put themselves to bed (or settle down for 'cockshut' as middle English has it) and so knowing where home is is something that they will appreciate.

However, chickens can frequently find themselves sharing their accommodation with some unwanted guests. As a chicken bluffer one home truth you can impart with confidence is that all the gentle, bucolic, and possibly sentimental books that have been published on the joys of chicken keeping will not major on the problem of vermin. There's no way round this: chickens and rats go together, and rats like nothing better than burrowing under static chicken houses, creating networks of tunnels that would be the envy of a Swiss mole.

One way to reduce the risk of your chicken's bijou residence being infiltrated by opportunistic rodents is to raise the henhouse so that there's a wind-blown, rat-unfriendly gap beneath it. The always resourceful Swiss also have a solution for this, building their grainstores and some chalets on stone 'mushrooms' or 'staddles' to prevent rats chewing through wooden stilts.

If the henhouse is high enough then the chickens who live in it can use the space underneath as a dust bath or somewhere to flop about and contemplate the world. However, if the run is reasonably high off the ground, so that getting in and out of it requires a fair bit of flapping, or precipitous jumping down, then a solid, reasonably gentle ramp will be required to aid chicken access.

Big, or heavy, chickens in particular can damage themselves if they have to leap too far too often.

SITTING PRETTY

Which brings the bluffer to perches. These will provide chickens with comfortable roosting spots and chicken bluffers with the opportunity to pass on some useful information.

Some athletic birds, particularly bantams, don't have an issue with heights, and will view rafters in a barn as entirely suitable spots to retire for the night, but some bigger, flight-phobic breeds (*eg* full-sized Buff Orpingtons) want perches nearer the ground, and something that might only be a couple of inches above it will fit the bill for them.

Completely round perches – broom handle shaped – aren't ideal, particularly for the bigger birds, as grabbing on to them and balancing is more of an effort. A rectangular section (for the technically minded bluffer, think about a '2 x 1') with the edge planed a bit so that chicken toes can wrap around them and gain 'perch purchase' represents the ideal.

Finally, before leaving the issue of housing, the question of security should be addressed. Given the opportunity other animals are happy to break in to the inner sanctum of any domestic chicken, so owners need to consider whether their fashionable chicken houses have floors that rats can't chew through and walls and roofs that can't be wrenched open by foxes and even badgers.

PRISON WORKS

It doesn't take an in-depth knowledge of chicken keeping to realise that making life difficult for these predators in the first place is a good idea.

This might mean creating a secure avian fortress with chicken-wire fencing surrounding it. To avoid predators calling a chicken keeper's bluff that wire needs to be robust and attached to solid posts or uprights that can't easily be dislodged.

A portable fruit cage pegged into the ground might look ideal, but if a fox can wrench it from the ground, upend it, and get to the birds inside then it's not fit for purpose. A permanent wire enclosure should provide more security, but any chicken bluffer worth their salt will point out that a henhouse needs to be completely enclosed. Foxes can effortlessly scale 6ft-plus garden fences, so won't be deterred by a roofless chicken enclosure, even if it's a couple of feet taller. Extreme security measures might include watchtowers, guards, spotlights and razor wire, but bluffers shouldn't get too carried away in this respect. A scaled-down Stalag Luft III (for imprisoned flyers) is not a realistic prospect. The truth is that there is no such thing as a completely predator-proof chicken house.

Bluffers should emphasise this, and also bear in mind another security consideration which might be thought of as self-evident but is often overlooked.

Although it is tempting to use chicken wire with fairly big holes in it, because it's easier to work and generally cheaper to erect, if the gaps are big enough for wild birds like sparrows and tits to break in on a feeding raid then baby chicks can get out. This isn't always good for their health.

BREAKING IN

Back to foxes, badgers and even pet dogs, who might view pet chickens as squeaky toys that can be played with until they stop squeaking. If they can't go up and over, they may well try going down. Foxes and badgers live in burrows, and are quite happy to tunnel their way towards domestic chickens. It's almost in their job descriptions to try.

Foxes and badgers live in burrows, and are quite happy to tunnel their way towards domestic chickens. It's almost in their job descriptions to try.

The answer is to dig a 2ft-deep trench around the run and bury the chicken-wire fencing into this, bending the wire outwards into an L shape. Unless predators are particularly industrious and dig really deep from the start of their excavations they will encounter the fencing rather than the chickens it is protecting.

All this stuff is irrelevant if the birds are free range and have the pleasure of wandering about discovering the wider world. A chicken philosopher rather than a bluffer might conclude that if they had a nice time doing this before meeting something vicious, hungry and possessing a sharp set of teeth, then the former made the latter worth it, although an owner discovering a sad pile of feathers might disagree. So might the chicken. Discuss.

EATING DISORDERS

People generally feed their birds twice a day, so they get breakfast and supper, and frequently something in between courtesy of indulgent owners. As a chicken bluffer you can make some sage observations about comestibles.

First off, food such as grain shouldn't just be chucked around willy nilly. Hens like nothing better than to scratch and forage around, but anything left is a magnet to rodents and wild birds that can carry unwanted diseases. Ditto the ground. If chickens live in confined space the ground can be a repository for all sorts of bugs, so encouraging birds to forage around in this is not a good idea.

Which, fascinatingly, leads to the subject of feed bowls. You might think that any old receptacle will do for putting food into, but no.

Chickens not only peck at food, they also scratch at it with their clawed feet, and will easily and inevitably upend lightweight food bowls. So proper feed troughs or relatively shallow dog-bowl style receptacles are the things to have.

Having somewhere to put them when the birds have retired for the night is also a must, because when chickens go to bed other nocturnal creatures can come out to help themselves to the leftovers.

Chickens are naturally omnivorous (a fact an avian bluffer will be only too happy to pass on) and their carnivorous tendencies mean that they will happily wolf down bits of expired mouse or entirely live frogs. As you know, they have cannibalistic tendencies if the circumstances are right (or, more accurately, wrong). Owners should keep a weather eye out for stuff they might like to eat but wouldn't do them any good if they did, and that includes each other. Foods to avoid – and here's an excellent, bluffable fact – includes rhubarb, which is poisonous to chickens, although that won't prevent them having a go at eating it anyway.

SCRAPPAGE SCHEME

Which brings readers who might claim to have moderate standards of neatness to kitchen scraps. Discarded or unwanted vegetable matter will often go down well with chickens, although owners may be disappointed at the lack of chicken enthusiasm when being fed with raw, unyielding sprout leaves, broccoli stalks and carrot peelings. Chickens have culinary preferences, and are more likely to welcome a range of foodstuffs if they've been cooked.

Sweetcorn will usually be very well received.

Chickens more or less regard this as the food of the gods, and will often eat it even when they're ailing and have forsworn almost everything else. This is a piece of information any self-respecting chicken bluffer ought to be familiar with.

Do chickens eat bread? Of course, and if you let them eat cake they will, but this doesn't mean that you should.

Bread has a fermenting quality which when it reaches a chicken's crop (the bulge at the base of the bird's neck where food starts its nutrient-extracting journey through the chicken) can set up a nasty reaction that can make the animal ill.

Compact and infrequent offerings of bread are probably OK, but anyone who feeds their birds on nothing else isn't doing them any favours.

Mixed corn and layer's pellets are what chickens should be fed on most of the time.

DRINKING HABITS

When chickens feel like a drink, they like nothing better than to drink from fetid puddles, the flavour of which can be further enhanced by said chickens wallowing in something unspeakable and then splashing happily about in the water before drinking it.

There is nothing owners can do about this. It's not nice, but it's part of chicken life. All that owners can do is make sure plenty of clean water is on tap, which means daily replenishing of the drinking apparatus and keeping this clean with regular scrubbing.

This sounds simple (if boring), but as a chicken bluffer you will know better. Some drinkers are resistant to being kept clean. Small plastic items with thin necks and bulb-like water compartments are to be avoided because it's almost impossible to get a scrubbing brush inside them to remove the filth.

CHICKEN CHARACTERS

As a seasoned chicken-keeping bluffer one of the things you will frequently be asked is whether chickens have personalities.

'They're all the same, aren't they?' is commonly the position of fowl-free antagonists, usually couched in a tone that leaves you in no doubt that this is a statement rather than a genuine question.

The reasons for this opinion often come down to an obdurate insistence that all chickens look and act the same. With little heads and even littler brains, so the theory goes, there isn't much room for the personality bit when that cognitive marshmallow is working overtime on less than cerebrally demanding challenges such as breathing, eating, scratching about, laying eggs and going 'cluck'.

Bluffers might concede that it's true that chickens have small brains when compared to those of human beings, dolphins and elephants, but the birds do have a surprisingly disparate world view, and as for looking the same, this is complete gobbledy-garbage. A crowd of people at a football match might adopt identical scarves and team strips and merchandise but look closely and you're confronted with infinite variety of ages, shapes, sizes and lung capacities. Which is another way of saying that chickens are endlessly individualistic. A Frizzle bantam, with its apparently permed curly feathers which stands about 8in tall isn't going to be confused with a Tuzo game bird, which has the look of a scaled-down dinosaur with showy exotic feather colours and a belligerent stance.

The personalities of these birds will be different too. Some chickens are entirely in your face, while others are self-effacing. Chickens can be scatty, fastidious, controlling, cowed, demented and staid. Just like people.

So, as a bluffer among poultry experts, being told/asked if chickens are all the same will afford you the pleasure of being magnanimous and condescending at the same time, because you know better. And since you know better, what could be more satisfying than offering up some examples of chicken personality types, using the following handy guide, starting with the female of the species.

CHIEF CHICKEN

This is the hen which is in charge of all the others. Chief chickens generally aren't the biggest, strongest or most aggressive birds, but usually they're the most senior. These animals are absolute rulers in cockerel-free henhouses, and in charge most of the time in those with a token avian bloke on the premises.

Chief chickens demand respect from those further down the pecking order, and if this isn't forthcoming will dish out selective justice, usually by pecking.

The less comfortable in her own feathers a chief chicken is the more likely she will resort to violence. Some birds have more natural authority than others, so will march to the front of the queue at mealtimes and quell any dissent with a 'don't mess with me' basilisk stare.

Some birds have more natural authority than others, so will march to the front of the queue at mealtimes and quell any dissent with a 'don't mess with me' basilisk stare.

SOCIAL CLIMBER

When time's winged chariot carries off a chief chicken, there will be another one directly behind her in the queue waiting to take over. As they've diligently climbed the social ladder over the years some of these birds will have been very keen to ensure that those on the rungs below are acutely aware of where things stand. This can mean an 'I'm bigger than you' approach to social interaction even when they aren't.

Some chickens will effortlessly rise through the ranks of domestic fowl society by reason of birth (much like human society). If you have a flock of mixed birds that have been sired by a particularly dazzling cockerel and reared by one of his favoured lady friends, those chicks can often be on a fast track to superior social status, overtaking those without such an illustrious genetic heritage.

THE BULLY

As with human social stratification, some of the worst behaviour in the chicken house can be found very nearly at the bottom of the social heap. Some birds, particularly younger ones, are wall-eyed, feather-wrenching little sadists, doling out nastiness to one or two hens that have the misfortune to be at the very bottom of the pecking order pile.

Some of these adolescent thugs have learned their behaviour while finding themselves under the claw of another more senior chicken malcontent, and have adopted an 'if I'm doing it to you, nobody's doing it to me' approach found in the darker recesses of our school playgrounds, offices and organisations.

Others just enjoy it, and will sometimes take this 'do you want some, then?' approach to the very top of the pecking order, which isn't much fun for the rest of the flock.

THE OBSERVER

Some birds develop strategies so that life is lived more on their terms, and this certainly applies to chickens who observe what's going on before piling in. Some big, heavy-breed chickens, such as Brahmas, seem to have a particular facility for this wait and see approach.

When there's food around these birds will hang back then make surgical strikes on the scoff once their more assertive cohorts have had their fill. Observer chickens also seem to have worked out that human beings are suckers for friendly chicken pursuing, a bit of eye contact and a look that says 'I might eat out of your hand if you get me some more grub.'

THE STALKER

Even the most pragmatic chicken bluffer is not immune to inter-species manipulation, and some hens have learned that being in their owners' faces is a very good conduit to extra food.

If your birds are free range, watch out for the one that will always run up to you whenever you venture into the garden and position themselves so that you're not quite tripping over them. They will then assume the role of feathery shadow as you move from flowerbed to flowerbed. Try digging a hole and they will stand exactly where you want to plunge the spade. Shoo them away and manage to dig a hole and as soon as you stop they will jump into it and murder any earthworms they can find.

These animals have worked out that being a bloody nuisance gets results, which means that if you have the temerity to ignore them and not provide copious treats from the food bin, they will wait until you've stopped moving then stand on your foot and give you a dirty look that says 'if you want me to bog off for at least two minutes, feed me, you dolt'.

THE EARTH MOTHER

Some hens will spend their entire lives without the merest hint that they would like to mother the next generation of chickens.

Others will go broody at the slightest provocation. This process can turn normally mild-mannered animals into hot, feathery, hormonal bundles of fury.

'(Some hens) with eggs will display a touching determination to protect them, lunging at any human fingers that come near, like enraged teapot cozies with beaks.'

Some will attempt to brood the straw in their nest boxes, apparently unaware that the process of incubation requires eggs. Others with eggs will display a touching determination to protect them, lunging at any human fingers that come near, like enraged teapot cozies with beaks.

Should any chicks appear, an earth mother hen will protect them with a puffed-up passion, showing them how to forage hours after they've emerged from their eggs, and carrying on as mentor and protector until the animals stuffing themselves under her wings are very nearly as big as she is.

Then quite suddenly she will forget who they are and will chase them off. They will look a bit bemused but not unduly hurt and inside a few hours will forget about her too.

THE LATCHKEY MOTHER

Some chickens suffused with the parental urge would be better off sticking to laying eggs rather than hatching things from them.

If these birds were human beings you get the impression that they'd be out clubbing most nights, getting smashed on alcopops and smoking themselves silly.

These birds will wander from the nest, forget that their eggs need to be kept warm so that their embryonic offspring are frozen stiff (or stiffs).

Should any chick survive this negligence, making it to adulthood will be a lottery, because mum will barely notice it, showing little interest in its well-being, and perhaps carelessly treading on it for good measure. Inevitably, such unsuitable mother hens will get the breeding urge every summer.

It's very hard to spot a latchkey mother hen, but you're less likely to see dereliction of child-raising duty from breeds that are known to be good parents (such as Silkies and Cochins, and among the bigger, heavier breeds Orpingtons, Brahmas and Sussex chickens). Warren, or hybrid birds, the sort of brown, egg-laying and meat-producing machines beloved of modern farming, are designed not to go broody, as when that happens they stop laying and are no longer productive, but sometimes there can be a genetic rebellion leading to the mothering instinct kicking in. Without being in any way judgemental, this commitment can sometimes be compromised by wavering hormones and chronic absent-mindedness.

THE FEATHERBRAIN

Some chickens go through their entire lives pottering about quietly, but others fling themselves at each day with demented intensity. These birds will have hysterics when a cat emerges from a bush (which is very empowering for the cat). They will stare at the middle distance then for no apparent reason shriek. This will be followed by a crazed flapping and running about drama which goes on until the animal forgets what the problem was and goes back to looking vacant or pecking about. Featherbrained chickens tend to be younger birds, and sometimes reach their middle years and dotages and decide that such behaviour is unbecoming and will change from idiocy to irritability.

THE GRUMP

Some grumpy chickens begin their lives in the featherbrain category, but will suddenly decide that life would be much more fun if they did a lot of sitting about and eyeballing everyone else with ill intent. Such animals have the demeanour of the sort of wizened, crinoline-wearing old ladies who appeared in Dickens's novels for the sole plot device that he needed someone to be nasty to orphans.

THE ALPHA MALE

Some cockerels are showy male tarts, strutting about looking glossy, pleased with themselves and resembling the chicken world's answer to the androgynous male models who pout and preen in masculine TV perfume commercials every Christmas. (Think *'Parfum de poulailler pour homme'*.)

Such animals aren't overburdened with self-doubt. They know they're gorgeous and, yes, they are God's gift to avian females thank you very much. None of this is particularly edifying or laudable, but such behaviour is a fact of life in many henhouses.

THE BETA MALE

Aggro-prone, noisy, unproductive creatures, who limped from the egg into a life of runtish frustration. He will be the bird who thinks he has something to prove who has absolutely no chance of proving it.

Given the opportunity he would square up to a sibling who could flatten him, or chase after female chickens who in some cases could flatten him too, and are less than keen on his liverish designs on them.

The only thing keeping this bird's mental health more or less on an even keel is a relentless egotism that will never be dented by the facts.

NAPOLEON COMPLEX CHICKEN

Some bantam cockerels definitely have little man syndrome. They might have the fighting weight of a bag of sugar, but that won't prevent a frazzled Frizzle or a psychotic Polish Crested Bantam from attacking the hand that feeds them, even if that hand is attached to a 7ft-tall professional rugby player with anger management issues.

This sort of behaviour tends to be at its worst during the summer, when chicken machismo is on the rise, and these diminutive feathered aggressors think every day is Fight Club day.

Even gentlemen chickens are driven by base hormonal urges which can lead to bad behaviour involving beak pointing, dancing and clucking. A posse of lady chickens will rush to the spot and begin foraging, wondering why there is nothing to find.

THE GENT

Some cockerels are genuinely solicitous of their girlfriends' needs and wishes. If confronted with a predator who can easily dispatch them (fox, badger, dog, etc) they will make a heroic, usually pointless sacrifice and go into a battle they're not going to win (although Alpha, Beta and Napoleon Complex chickens would probably do the same).

At mealtimes they will ensure that their favourite ladies don't go without (sometimes at the expense of less favoured partners, who will find gentlemanly behaviour in short supply).

When out and about having a garden forage this can look quite touching. An avian Sir Galahad will spot a special edible treat, such as a slug, and will commence a little food dance, and using his beak to point at the slug, who will probably be making a very slow run for it.

This display will attract the attention of any nearby hens who will rush over and, assuming the slug has travelled all of an inch and a half, devour it.

So far so gallant, but even gentlemen chickens are driven by base hormonal urges which can lead to bad behaviour involving beak pointing, dancing and clucking. A posse of lady chickens will rush to the spot and begin foraging, wondering why there is nothing to find.

This will be in a heads down, bottoms up posture, which means the cockerel, who has been lying about the food, which is non-existent, can jump one of the foragers from behind for some distinctly ungentlemanly, and indeed unwanted, love action. The utter cad.

This behaviour might be disappointing in a moral sense and will undermine the otherwise gentlemanly demeanour of some birds, but for a chicken bluffer, knowing about this sort of cheating is a real conversation stopper.

THE COWARD

Many cockerels, consumed with an irrational jealousy, see their owners as love rivals. This will result in a lot of chest puffing and charging about.

Owners will approach their chicken runs bearing edible gifts to be greeted by a testosterone-addled chap chicken flinging himself at the wire in a 'come here if you clucking want some then!' kind of way, only to retreat if you do. This is because the bird in question thinks he's a hard man, but in fact he's a total wuss.

Encounter this animal on neutral territory and such a bird might engage in some martial arts-style leaps and kicks, but there will be no physical contact because all this thrusting about will be taking place about 10ft away from you.

A cowardly cockerel will have worked out that ultimately you are bigger and stronger than he is and will have taken a pragmatic approach to having a 'keep off my birds' stand-off akin to a small nation with a couple of minesweepers engaging in military exercises just outside the territorial waters of the US Pacific Fleet. If you call the animal's bluff by walking towards him he will probably run away squawking.

Sometimes self-preservation will be forgotten, the red mist will descend on the wimpiest cockerel's brain and there might be accidental physical contact – perhaps some inept jumping up and down on one of your Wellington boots, before he remembers that you are bigger and scarier than he is and runs away.

An effective response to such behaviour might be

described as tough love, with the love bit being the most important. This is metered out by tracking down and capturing the bird, then cuddling him, tickling him under the wattles and giving him a birdy shoulder massage.

This will result in much confusion, as the animal wrestles with the urge to peck you to death and the fact that on a sensory level what you are doing feels rather nice. There will be a struggle along the lines of 'just wait until you've put me down! You're really going to pay for this now sunshine!' and 'actually, I've got a bit of an itch at the base of my neck just *there*. Oooh, left a bit, right a bit . . . aaaah . . . you've got it.'

Once released the bird will usually be so confused that he will retreat to a safe distance, still shaking his fist (if he had one) and uttering bloodcurdling avian oaths.

THE PSYCHOPATH

For some cockerels such blandishments are useless, and the consequences of trying them out are likely to be unsatisfactory and painful.

These birds are genuinely mad as hell, you are a love rival and they're going to sort you out. These psychopath chickens will not be swayed by warm words and a back rub. They do hate your guts and they do want to kill you. Stout clothing, boots and gloves and hockey masks are probably the best protection against their demented attentions.

Chickens engage in a great deal of scrabbling about so can suffer cut, scratch and infection-related poultry podiatry problems (try saying that more than three times in swift succession).

OFF-COLOUR FOWL

Welcome to a jolly compendium of avian ailments. No book about chickens would be complete without a list of chronic infirmities that can make domestic fowl feel foul.

Knowing a bit about these often-terminal conditions can bring any chicken bluffer undreamed of kudos with hardened chicken experts and newbies alike, so here's a hit parade of chicken diseases just for you.

BUMBLE FOOT

This is a nasty affliction that causes a feathery patient's foot to swell up and its pad to redden. Often you can find a scab-like circle at the base of the affected foot. Chickens engage in a great deal of scrabbling about so can suffer cut, scratch and infection-related poultry podiatry problems (try saying that more than three times in swift succession).

The condition's official name is *plantar pododermatitis* (also known as 'footpad dermatitis') and is caused by an invasion of bacteria, such as *staphylococcus* – a name every chicken bluffer should remember – that causes an infection resulting in a warm, puss-filled abscess.

Clean, dry litter, perches that are less than 18in from the ground, and a balanced diet so the birds don't get fat will all reduce the risk of this sometimes fatal ailment taking hold.

Treatment suggestions vary from Epsom salt baths to honey poultices (honey is a hydroscopic substance so can draw out liquid), antibiotics and bandaging to surgical interventions. That's all you really need to know for bumble foot bluffing purposes.

COCCIDIOSIS

Coccidiosis is a parasite that attacks the gut wall of chickens. Loose droppings are often the first sign of trouble. Hunched, listless birds who've given up eating are prime suspects. Especially when they turn down fresh sweetcorn.

The parasite starts off in a sort of egg known as an unsporulated oocyst, which can live in the ground for years before going to work on a hen's intestines once it's been eaten, changing form and technical names as it does its worst. It's not unlike the plot of a sci-fi horror film.

Treatment involves specialist medication, antibiotics, and giving off-colour birds probiotics and multivitamins is also recommended.

Specialist disinfectants will often be needed to scrub out henhouses, as coccidiosis scoffs at normal domestic products. Infected birds should be kept in clean, dry and damp-free conditions. But then so should normal healthy birds.

EGG BINDING

This eye-watering condition is the result of eggs getting stuck in transit inside chickens before they're expelled.

Obese hens are more prone to it than their slimmer siblings, a lack of calcium or other nutrients can be causes, as can oviduct* infections of young birds starting to lay before they've fully matured. Sometimes very big or oddly shaped eggs can get stuck.

Symptoms include general lethargy, loss of appetite, and in some cases a rocking, waddling gait, which sounds diverting, but isn't, as egg binding can cause serious distress,

Suggested treatments include rubber gloves, lubricants and specialist, digital rootling about in a rather sensitive place. It's not necessary to go into specifics in further detail. Giving the patient a warm bath in a sink or plastic container for fifteen minutes or so, followed by a vigorous towelling down is another option, as is holding the chicken's backside over a pan of hot water or aiming a hair dryer at it.

EGG PERITONITIS

Another fundament-affecting condition with plenty of 'yuk factor' potential, this is caused when the lining of

* See the chapter 'Eggsplanation'.

the abdomen is inflamed thanks to a bacterial infection, and yolk goes into the abdominal cavity and sets up an infection.

A buildup of fluid can result in a bulging, pendulous abdomen. Usually it's too late to do much when the symptoms present themselves. In the past repeated courses of antibiotics have been used to keep the condition at bay, but given antibiotic resistance worries, this might not be such a good idea, and anyway the condition is more often than not fatal. Sorry.

This is all a bit gloomy, but on a more cheerful note implants have been developed to prevent laying so that a patient's immune system can get to work on the infection.

FOWL POX

This unlovely condition isn't always a killer, but it's painful and hard to shift.

Anyway, welcome to the world of a virus that can make chickens, pigeons and even parrots look warty. The virus can be spread by wild birds and biting insects.

Bluffers with strong constitutions need to know that there are wet and dry versions of the condition, and that the latter can spread to a victim's windpipe and can be fatal.

There isn't a cure as such, although antibiotics have been used to keep secondary bacteria at bay. Buying new birds from reputable breeders to reduce the risk of introducing the disease is important too. One sensible tip is that new stock should be quarantined for 21 days before being officially introduced to everyone else to see if any nasty symptoms emerge.

IMPACTED CROP

The crop can be found at the base of the neck. It's an area that contains grit that the bird uses to mash up the food before despatching it to the gut.

Sometimes food gets stuck in this area, either because the bird has eaten something it shouldn't (like plastic or string), or there can be issues with muscles not contracting when they should.

A lack of grain and eating too much long grass can also set up the problem, which sometimes requires surgery to extract the offending gloop.

Any chicken bluffer worth their A–Z of fowl diseases will at this point namecheck sour crop, a related condition, where the crop is compromised by a yeast infection. Birds become listless, disinterested in eating, have fluid-filled crops and breath that smells like a home brewing kit.

The prognosis isn't great and treatment options sketchy, but the helpful chickenvet.co.uk website recommends a prebiotic called Beryl's Friendly Bacteria to restore the crop's normal, benign bug culture if antibiotics have been used.

MAREK'S DISEASE

There's a poor-taste joke that anyone who kisses a budgie risks contracting 'chirpees', but Marek's, a member of the herpes family, isn't funny at all – as any chicken that gets it is effectively doomed.

Chicks and pullets (teenaged chickens) are particularly susceptible and once in the bloodstream the virus will compromise the immune system, complications ranging from internal tumours to nerve damage follow.

The virus can be spread in feather dust, so is easily transferred, particularly if chicken housing isn't kept clean. The usual caveats about cleanliness and care when introducing new birds into a flock apply here.

PROLAPSE

This is a problem when older chickens in particular let it all hang out, and not in a good way.

When laying eggs a hen's reproductive track is turned inside out but when prolapsed it doesn't snap back into place again. Former battery hens who've been mass-producing eggs for their entire adult lives and have stretched tubes are particularly at risk.

Vets will sometimes prod the distended innards back into place, which owners can also try, and yes we're back in the land of rubber gloves and lubricating unguents. Cleaning the area by running warm water over it but not otherwise actively doing anything can also help.

Anyone hoping that nature will take its course more slowly will need to isolate the patient, keep her somewhere dark most of the time (light triggers the egg-laying urge), and offer soft food and plenty of fluids. If in doubt, don't bluff. Visit the vet.

SCALY LEG

A condition caused by mites called *knemidocoptes mutans* that get under the scales on chickens' legs and toes causing them to lift and become irritated. Lameness can follow.

Treatments include washing the infected areas with warm water infused with baby shampoo. This can be worked in with a soft tooth or pastry brush. After that coat the infected areas in Vaseline as this will smother the horrid little mites. A full cure can take upwards of a year but regular treatment should see improvement pretty quickly, and should be done in tandem with henhouse disinfection.

TOE BALLS

These are rock hard beads of mud and guano that build up on the end of a chicken's feet. Prolonged, repeated dunking of chicken feet in warm water and gentle toe ball working is recommended. That and keeping an eye on birds so that things like this don't build up and become big problems before they're noticed.

VENT GLEET

A disease any chicken bluffer will be glad to name-check, 'vent gleet' will result in birdy bottoms caked with a nasty yellowish paste and a rather foul smell.

There's some debate about whether the condition is caused by a herpes virus or thrush-based infection.

The British Hen Welfare Trust recommends sacrificing a washing-up bowl, half filling it with warm water, adding two tablespoonfuls of Epsom salts followed by the malodorous patient.

Keep her there for ten minutes, during which time she might well fall asleep. After that wake her, dry her, apply Canesten cream to her vent or backside area, and just inside it (welcome to the joys of chicken keeping) and a little apple cider vinegar to her drinking water and provide some probiotic once a day.

Repeat this procedure every couple of days for a week. Alternatively, take her to the vet, which is the bluffer's advice nearly every time one of the above lurgies show their malign (and expensive) presence.

CHICKENS IN POPULAR CULTURE

It's said that human beings are never more than 6ft away from a rat, but given their success as a domesticated species, and their more recent elevation to designer pet status, this might soon be true of chickens as well. Perhaps that explains why the birds have become such a feature of popular culture.

This offers a rich seam of bluffing possibilities which every chicken aficionado needs to know, and which could be subtitled 'how chickens took over the world'.

FOOD FOR THOUGHT

Even chickens which have led a largely unremarkable life as broilers have left their mark on British culture. Consider chicken tikka masala, a dish of Indian origin that some have dubbed Britain's national food of choice thanks to its immense popularity.

The masala is said to have been cooked up to stimulate a Scottish palate which had been blunted by uninspiring fast food, although there's a counter argument suggesting that chicken tikka masala is actually an authentic Indian recipe and several hundred years old. This gives you plenty of opportunity to engage in 'was it or was it not?' bluffing.

You, of course, will be aware of the interesting culinary digression that Glasgow's Shish Mahal restaurant lays claim to be recognised as the place where chicken tikka masala was invented by head chef Ali Ahmed Asham with a tin of tomato soup, some cream and a handful of chillies and spices in the 1970s (aka the decade that taste forgot). You might add that the restaurant attempted to secure the dish the same legal protection as other 'regionally designated food', using 'Arbroath Smokies' as an example of what they meant.

A HEAD START

No self-respecting chicken-keeping bluffer can fail to know about Mike the 'Headless Chicken' from Fruita, in Colorado. Mike was a regular, five-month-old farmyard Wyanadotte rooster belonging to a farmer called Lloyd Olsen. Given Mike's inability to lay eggs and apparent indifference to procreation, he was preordained for a life involving hatching, scratching and despatching.

In 1945 farmer Olsen decided to expedite the last of these activities with an axe, and decapitated poor Mike.

Instead of keeling over Mike staggered about and then refused to die. Soon the headless bird was having a go at preening – without much success, and scratching around as usual. The wince–inducing truth was that Mike's jugular vein had sealed so he didn't bleed to death and there was enough residual brain stem for him to keep on chickening around. He'd also retained one ear, so could listen to the world in mono even if he could no longer see it.

The morning following his beheading Mike was found asleep rather than deceased. Since conventional eating was out of the question, Olsen began giving the bird fluids and nutrients with an eye dropper, and was soon exhibiting the unfortunate beast as 'The Headless Wonder Chicken'.

Apparently avian decapitation was clearly a conduit to the American dream, and Mike was soon on tour. His owner described Mike as 'a robust chicken' – a fine specimen of a chicken except for not having a head.

Celebrity beckoned, and the headless bird was soon touring the USA, charging punters in San Diego, Los Angeles and New York 25 cents each to see Mike. He was eventually insured for $10,000, made the pages of *Time* and *Life* magazines and even the *Guinness Book of Records*. The celebrated chicken lasted another year or so, only to expire in a motel room in Arizona, falling victim to a choking fit.

GOOD CAREER MOVE

Exit Mike, but he left behind a legacy in Fruita which still holds annual 'Mike the Headless Chicken days', involving blindfolded participants racing each other over five kilometres in suitable costumes and chicken-themed dancing, where strapping rednecks 'Do the Funky Chicken'. At least one song has been written about Mike, with the questionable title of 'The Cluck Stops Here'.

CLOSER TO HOME

The Isle of Man is a good place to be if you're a fan of infamously perilous motorcycle racing, horse-drawn trams, tailless Manx cats and home rule. It can also be a positively lethal environment if you're a chicken. This is thanks to an occasionally resurrected centuries-old island tradition called Laa'l Catreeney, otherwise known as St Catherine's Day, which takes place on 6 December, and involves a mock funeral for a chicken that is bumped off, decapitated and then deprived of its feet.

Young men would then parade various bits of the dear departed around the island, singing:

'Catherine's hen is dead,
You take the head,
And I'll take the feet,
And we'll put them under the ground.'

The funeral follows, and the chicken's head and feet are buried. The feathers are kept for 'good luck', something their former owner had clearly run out of. Having got this frankly sinister and rather pagan-sounding ceremony out of the way mourners retire to a pub where what is left of the sacrificial chicken ends up on the menu, while everyone celebrates its life by getting blind drunk on something called 'jough'. It isn't necessary to know what 'jough' is, except that it comes in liquid form and is more accurately known as 'chicken's revenge'.

Those who drink it are said to have 'plucked a feather of the hen'. Given the island's fame as a centre for creative accounting, these days such an activity would presumably be tax deductable.

BOOKWORMS

Unsurprisingly chickens have found their way into plenty of books and while titles such as *Beautiful Yetta the Yiddish Chicken*, the 2010 children's bestseller by Daniel and Jill Pinkwater, might instantly spring to mind as one of the most recently successful examples of 'chick-lit', there are other chicken-related storybooks which have been familiar to children for centuries in various forms.

The Little Red Hen has been around since the mid 19th century and tells a tale of the importance of the work ethic and the dangers of sloth. Generations of children have failed to escape this morality tale of an industrious chicken who finds a grain of wheat, asks other farmyard animals to help her plant it, but finds that they are all too lazy to help her. So she makes a loaf of bread from the resultant harvest, and refuses to share it with all the animals who wouldn't help her earlier, but were happy to eat the results of her labours.

The story was used for a 1934 Disney cartoon, *The Wise Little Hen*, which is where Donald Duck made his debut, and a character called Tom Turkey was 'eliminated during the story development'.

The story was also the subject of a 1970s radio monologue by actor-turned-American president Ronald Reagan (before he entered the White House), and also featured in the song 'Little Red Rooster' by the blues singer Willie Dixon.

Few would argue that *The Little Red Hen* is a giant figure in children's chicken literature (although some might venture to suggest that she's also a sanctimonious, proselytising one as well).

Chicken Little (aka *Henny Penny* and *Chicken-Licken*) is another children's story that began life as an oral folk tale about a young chicken who becomes convinced that the sky is falling in. It became a surprisingly successful Disney computer-animated film in 2005. The moral of the original story is to have courage, even when it feels like the sky is crashing down. In Henny Penny's case the moral is slightly different in that life doesn't always work out quite as the chicken had hoped, and hungry foxes are never to be trusted. But that's just one of a number of different versions. Take your pick.

POULTRY IN MOTION

Chickens make perfect cartoon characters, as animators for the likes of Disney and Hanna-Barbera's *Tom & Jerry* (which featured a cat-bashing rooster) discovered and it was only a matter of time before the birds got their own movie in the shape of Bristol-based Aardman

Animations' *Chicken Run*, which used stop-frame animated puppets.

Released in 2000 by the creators of the *Wallace and Gromit* series, *Chicken Run* was Aardman's first feature film and apparently remains the highest-grossing animated film ever made, beating the likes of Wes Anderson's *Fantastic Mr Fox*.

Chicken Run is that rare thing, a feminist fable (with few exceptions all the male characters are incompetent twerps and cowards), with a plot in part inspired by the *Dambusters* movie, a black and white, stiff upper lip celebration of British pluck and in particular *The Great Escape* from 1963, a film that featured Steve McQueen and his fence-jumping motorcycle exploits, which *Chicken Run* cheerfully parodies. It's only slightly ironic that the serially unreconstructed Australian Mel Gibson was one of *Chicken Run*'s stars, along with Imelda Staunton, Jane Horrocks and Miranda Richardson.

Each minute of footage took about a week to film, which is why the shoot took 18 months, and that followed a two and a half year character development and story boarding exercise.

BEYOND FAMOUS

Since this is an age of celebrities, real, manufactured and imagined, it would be remiss of a committed chicken bluffer not to know the names of a few public personalities who are private hen keepers.

In no particular order we have shy, retiring **Jeremy Clarkson**, who has been known to keep chickens at his home in Oxfordshire, where it's possible they became passive smokers. **John Cleese** has even been known to let members of his flock into his kitchen – a YouTube search should reveal the footage. One of them was named Camilla, possibly in honour of the Duchess of Cornwall.

The actress **Jennifer Garner** is a chicken keeper, who for reasons entirely unconnected with self-publicity has been photographed walking a chicken on the end of a leash along a suburban street.

Older chicken bluffers may not be familiar with someone called **Kylie Jenner**, but she's a reality television 'personality' who appears in something called *Keeping Up With The Kardashians*, so is very, very famous. She owns a silkie cockerel called Eddie, and apparently has kissed the top of his head on Instagram, which is big news.

Television chef and scourge of the Turkey Twizzler **Jamie Oliver** has done for more than a few chickens in his professional life, but the man who no longer says 'pukka' when appearing on screen is also a chicken keeper. **Madonna** has kept the birds too, although whether she has since divorced them and launched legal proceedings for custody of their coop remains unclear.

Among the dear departed chicken keepers we have Hollywood actress and wife of Clark Gable, **Carole Lombard**, who had a pet cockerel named Edmund, who posed nervously with the star after she'd plonked him on to a rather high fence post.

Surrealist painter **Salvador Dali** tended to use animals as props, and was photographed emerging from a Parisian metro station with his pet anteater on the end of a lead. He also owned an ocelot called Babou, who presumably could have eaten Dali's pet bantam cockerel, a patient animal that put up with being photographed on the artist's shoulder, just inches from his carefully sculpted moustache.

Other chicken enthusiasts include **Julia Roberts**, **Jennifer Aniston**, **Reese Witherspoon**, **Martha Stewart**, and **Tori Spelling**, the latter an American actress who not only keeps chickens but also designs clothes for them. As yet there is no evidence of an infamously revealing designer creation held together by safety pins and adapted for chickens, but the next best thing is the memorable gown worn by actress and organic farmer **Elizabeth Hurley** whose farmstead in Gloucestershire quite naturally contains a number of chickens. La Hurley might no longer be on newspaper front pages every other day, but by dint of her inclusion in this modest volume she will now have the honour of being at the apex of chicken bluffery, which is an achievement in itself.

All this might be perceived by non-chicken fanciers as terribly parochial and unfashionable, but the people and chickens involved don't care and only a cynic would fail to be impressed by the beguiling charm of these human/chicken love-ins. Go to a chicken show and be prepared to leave your doubts at the door.

BEST IN SHOW

Some chickens have the haughty strut of catwalk models, and with good reason, because that's what they are.

These birds haven't been bred to provide eggs, be eaten or add kinetic decorative features to a suburban garden. No, they've been put on Planet Earth to be exhibited at chicken shows, and that means they live a rather rarefied life with somewhat reclusive overtones.

No wild parties or hedonistic lifestyles for these glittering examples of avian pulchritude. To keep young and beautiful a show chicken is likely to be kept away from mud, muck, other chickens and the potential to injure itself or in some way damage the feathery body beautiful.

Many will live indoors, spending their days in scrupulously clean chicken penthouses. This doesn't mean they aren't cared for or indeed loved. Visit a chicken show and you will be in the company of doting owners, but the advice from the Poultry Club of Great Britain about what is expected of a show chicken is stern and uncompromising, emphasising that preparation has to start months before a show.

It warns that some breeds have to be hatched about five months before being exhibited to be at their absolute peak, but large fowl will take longer.

'Positive signs of health should be present in every potential show bird including a bright eye, red comb, dry nostrils, shiny feathers, a good weight, clean feathers under the tail and an alert active manner,' it says on the club's website.

There is a 'breed standard' which covers 100 points and will exclude otherwise stylish chickens with the wrong coloured legs. 'If your bird should have white legs (or plumage) do not feed maize in order to avoid a yellow tinge and avoid too much sunshine as white feathers will gain an undesirable straw tinge,' warns the club. As a chicken bluffer this means you now know the importance of making a chicken pale and interesting.

BEAUTY PARLOUR

Many show birds are extremely tame and laid-back, probably because they're handled a lot. Generally they are washed, shampooed and in some cases given leg waxes. As the Poultry Club puts it:

'The birds are dunked (up to the head) in warm water, lathered (brushing feathers away from the head), rinsed thoroughly and initially towel-dried and finished either in front of a fire or with a hair dryer. The birds will usually enjoy the experience if it is done carefully and thoughtfully.

'Every bird will need its feet and legs thoroughly scrubbed clean in warm water before showing. A cocktail stick or nail file can be used to very carefully and gently remove any remaining dirt from under the scales. Coconut oil, available from health food shops can be applied to the legs and will stop the dirt getting back under the scales,' advises the club.

Often many of these processes will have taken place several days before the show so that natural oils that make feathers shiny have a chance to re-assert themselves.

LOOK AT ME I'M WONDERFUL

Many show chickens do seem to enjoy being the centre of attention, and at shows are entirely phlegmatic as they're given a last-minute wash and brush or are peered at by hundreds of chicken fans.

Prior to judging the Poultry Club recommends using 'a little oil, Vaseline or hand cream rubbed into the comb, wattles and legs will spruce the bird up. A silk handkerchief is said to be good for improving a shine to the feathers.'

Judges are on the lookout for shape, colour, form and posture and an indefinable something that separates chicken contest A Listers from clucking X Factor wannabes.

There are chicken shows all over the country, which involve a good deal of travelling about for serial competitors and a good deal of work behind the scenes involving breeding, feeding and general preparation, and competition can be fierce. It can only be a matter of time before *Little Miss Sunshine* is re-released with an all chicken cast.

Anyone attending such an event, even if they're on an information-gathering exercise for chicken bluffing purposes, will almost certainly encounter a friendly, good-natured ambience and some frankly amazing-looking animals. There is also an entertaining and arcane selection of show classes, including junior categories, where tomorrow's hen keepers can demonstrate their chicken handling skills as well as the birds themselves.

All this might be perceived by non-chicken fanciers as terribly parochial and unfashionable, but the people and chickens involved don't care and only a cynic would fail to be impressed by the beguiling charm of these human/chicken love-ins. Go to a chicken show and be prepared to leave your doubts at the door.

EGGSPLANATION

In the annals of unusual facts with which to impress your friends, knowing how eggs are made might not be at the top of your list – but if you're holding yourself out as something of a chicken expert, perhaps they should be.

The thing is, eggs are a great deal more interesting than might be imagined, perhaps because we take them for granted seeing them as little more than nature's convenience food with entirely environmentally friendly biodegradable packaging.

However, we live in an age where it's very much 'on trend' to know where food comes from, and egg expertise, even if it is not terribly expert, will win you bluffer's brownie points because, although we know that eggs come from chickens (which inevitably come from eggs), these eggs have already been on a convoluted, albeit rapid, journey through the bird.

According to the real experts at the British Hen Welfare Trust, there is a sort of egg assembly line inside a female chicken, which begins with the left ovary, which is the only one that functions in a chicken – the sort of information that bluffers thrive on.

The ovary contains follicles of differing sizes and maturity. In this instance the follicle has nothing to do with hair, but is instead a fluid-filled sac that contains an immature egg, or 'oocyte'. When a follicle matures, it becomes known as a yolk or vitellus. Now the yolk exits the ovary and moves on into the oviduct, the passageway also known as the uterine tube. The first part of the oviduct is known as the 'infundibulum', and the egg hangs around in it for about a quarter of an hour. If the bird has been jumped on by a randy cockerel this is where the resultant fertilisation takes place. Anybody interested in chicken birth control should bear in mind that sperm can lurk here for up to a month. Keen chicken breeders, or indeed avian eugenicists with plans for developing specific chicken blood lines can be in for an unpleasant surprise and see their plans entirely de-railed if more than one cockerel has been 'at' the egg-producing chicken.

The next part of the oviduct is called the 'magnum' and it's where the egg remains for approximately three hours, while the white is generated. That part of the process behind it, the egg gets moving along the oviduct until it arrives at the isthmus where it spends about sixty minutes as membranes are formed round it. The egg will then spend a day in the uterus as the shell is formed. The final stage of an egg's journey takes place in the cloaca, where a hormone which the bluffer will commit to memory as 'arginine vasotocin' sets up uterine contractions, and with a bit of effort, often accompanied by some 'I want you to know I'm working hard at this' noises, the egg is laid.

EGG FIRST, OR CHICKEN?

If they're fertile and a broody chicken sits on them for around three weeks, chicks will usually result. Embryonic development will actually have started while the egg is still inside the adult bird.

Spare a thought for the broody hen. She's running a temperature and has possessive and rage-drenched hormones coursing through her. She will often have amassed a large clutch of eggs. Frequently she will have pinched those laid by other hens as surrogacy is very much part of chicken society.

After twelve days the embryo has developed into a chick, with its head pointing towards the fatter part of the egg, which contains an air cell. The tiny bird breaks into this and begins breathing on its own. Eventually it cracks a hole in the shell which it gradually expands until it breaks free, a process that can take anything between four and twelve hours.

This is usually such an exhausting process that the chick which has emerged wet, matted and dishevelled, will lie prone for several hours as it gives its metaphorical batteries time to recharge. Once that's done it will quickly become fluffy, and extremely active. In a day or so the new arrival and its brothers and sisters will be out and about with their mother learning about life. Mother hens will protect their offspring with single-minded dedication, and will insist that they stick close to her until over the space of little more than a day a hormonal switch will be thrown, she will cool down, forget that these gawky animals were ever anything to do with her and permanently disown them.

Her suddenly discarded children will find this unsettling for a day or two, following her around until they give up and forget that she was ever their mother. Soon they too will have heads filled with thoughts of food, sex, power struggles and laying eggs. Of course, eggs are where a chicken's world starts and also where *The Bluffer's Guide to Chicken Keeping* comes to an end, partly because knowing about how the domestic fowl life cycle works will confirm you as a true chicken bluffer, and because the age-old question of whether chickens or eggs came first remains unanswerable.

POSTSCRIPT

There's no point in pretending that you know everything about keeping chickens – nobody does (including the author). But if you've got this far and absorbed at least a modicum of the information and advice contained within these pages, then you will almost certainly know more than 99 per cent of the rest of the human race about how to join the many thousands of amateur chicken keepers. Most importantly, you will know how to pretend to know more about the subject than you actually do. What you now do with this information is up to you, but here's a suggestion: be confident about your new-found knowledge, see how far it takes you, but above all, have fun using it. You are now a fully-fledged expert in one of humankind's most valued and unique traditions – the suggestion, artfully disguised and always falling the right side of an outright lie, that you are a world authority on the matter at hand.

GLOSSARY

Araucana. Flighty breed of fancy chicken. A fast mover that lays greenish eggs, and often hides them.
Beak clipping. An unpleasant practice that's largely fallen out of use in Britain. Once used by some factory farmers to prevent caged birds injuring each other.
Candling. A way to check if an egg is fertile by shining a light through a tube behind it so that its contents are revealed.
Chicken joke? Here's one: What do you call a weird and scary chicken? Answer: a poultrygeist....
Ducklings. Broody chickens will sit on duck eggs, although the gestation is a week longer for ducks, whose habit of heading for water at an early age can alarm chicken surrogate parents.
Egg binding. When a bird's egg-laying tract goes on strike and an egg gets stuck. Gentle application of heat will often cure this.
Feathers. Thought to be an evolutionary development of scales – a nod to the reptilian past of birds.
Grit. Chickens need a gizzard full of the stuff to process

and digest food. In this context grit has nothing to do with perseverance or pluck. Just cluck.

Hay. This is not suitable for chicken bedding as it contains spores that can lead to breathing problems. Wood shavings are the best option for coop bedding, but make sure they're pine wood and avoid sawdust at all costs.

Kiev. As in chicken kiev. Allegedly originating from Russia circa 1912, this dish stuffed with garlic and parsley butter was a 1970s culinary staple.

Laying. Chickens don't lay eggs all year round. They stop when it's dark, when they're broody, ill or growing new feathers. Or just feeling bolshy.

Marans. A speckled breed of hen originating from France. Nothing to do with meringues, although they sound vaguely similar.

Night. Generally not a good place for fowl friends. A chicken that's out on the town after sunset is a sitting target for predators because it can't see in the dark.

Ovo-lacto. Branch of vegetarianism which allows consumption of eggs.

Poultry. A Middle English word dating back to the 1400s, although the spelling 'pulletrie' could be used.

Red mite. Not in this context anything to do with the strength of the former USSR's armed forces, but a blood-sucking parasite that can infest chicken houses.

Swimming. Although generally unenthusiastic about immersion in water, chickens are capable of a duck-like paddle if they fall in.

Teeth. Chicks have a section of bone on their beaks known as an egg tooth. This allows them to break out of their eggs and vanishes as they mature. Useful to point out when someone's using the expression 'as rare as hen's teeth'.

Uterus. Despite being an egg-laying animal, the female chicken does have a uterus. It's known as a shell gland. Top bluffer's insight.

Vent. Technical name for a chicken's backside. Sorry.

Wheezer. A slang term for the vent. Bit of a whiff.

X. There is nothing that begins with X that relates to chickens. Please move on, unless you happen to know of a couple of Leghorns called Xanda and Xerxes.

Yamato Gunkei. A large Japanese fancy fowl with a wrinkly comb and face. It looks as if it belongs in a tuxedo outside a nightclub. Inevitably nicknamed Tomato Gunk.

Z chromosome. Similar to the human X chromosome, but one that puts chickens at the end of the alphabet.

USEFUL LINKS

The Poultry Club of Great Britain
(repository of information on breed clubs, chicken shows and all sorts of other avian wisdom).
info@poultryclub.org

Andy Cawthray (practical information on husbandry, plants and poultry). thechickenstreet.wordpress.com

British Hen Welfare Trust
(national charity that re-homes commercial laying hens and encourages support for British free range eggs).
bhwt.org.uk